Maintenance, Replacement, and Reliability

Theory and Applications

MECHANICAL ENGINEERING
A Series of Textbooks and Reference Books

Founding Editor

L. L. Faulkner

*Columbus Division, Battelle Memorial Institute
and Department of Mechanical Engineering
The Ohio State University
Columbus, Ohio*

Maintenance, Replacement, and Reliability

Theory and Applications

Andrew K. S. Jardine
Albert H. C. Tsang

Taylor & Francis
Taylor & Francis Group
Boca Raton   London   New York

A CRC title, part of the Taylor & Francis imprint, a member of the
Taylor & Francis Group, the academic division of T&F Informa plc.

Published in 2006 by
CRC Press
Taylor & Francis Group
6000 Broken Sound Parkway NW, Suite 300
Boca Raton, FL 33487-2742

© 2006 by Taylor & Francis Group, LLC
CRC Press is an imprint of Taylor & Francis Group

No claim to original U.S. Government works
Printed in the United States of America on acid-free paper
10 9 8 7 6 5 4 3 2

International Standard Book Number-10: 0-8493-3966-9 (Hardcover)
International Standard Book Number-13: 978-0-8493-3966-0 (Hardcover)
Library of Congress Card Number 2005041893

Library of Congress Cataloging-in-Publication Data

Jardine, A. K. S. (Andrew Kennedy Skilling)
  Maintenance, replacement, and reliability : theory and applications / Andrew K.S. Jardine and Albert H.C. Tsang.
        p. cm -- (Mechanical engineering)
  Includes bibliographical references and index.
  ISBN 0-8493-3966-9
  1. Industrial equipment--Maintenance and repair--Mathematical models. 2. Reliability (Engineering)--Mathematical models. 3. Replacement of industrial equipment--Mathematical models. 4. Mathematical optimization. I. Tsang, Albert H. C. II. Title. III. Mechanical engineering (Marcel Dekker, Inc.)

TS192.J37 2005
658.2'02--dc22   2005041893

Taylor & Francis Group
is the Academic Division of T&F Informa plc.

**Visit the Taylor & Francis Web site at
http://www.taylorandfrancis.com**

**and the CRC Press Web site at
http://www.crcpress.com**

Dedication

To my wife Renee, daughter Charis, and son Alvin.

Albert H.C. Tsang

To BANAK (minus an A), their spouses and bairns (Cameron, Callum, Lachlan, and Meghan).

Andrew K.S. Jardine

Preface

The purpose of this book is to provide readers with the tools needed for making data-driven physical asset management decisions. It grew out of lectures given to undergraduate and postgraduate students of various engineering disciplines or operational research, and from continuing professional development courses for managers, professionals, and engineers interested in decision analysis of maintenance and asset management. The contents have been used to support such courses, conducted by both authors individually or together, on numerous occasions in different parts of the world over the years. The presentation of the decision models discussed in Chapters 2 through 5 follows a structure comprising: Statement of Problem, Construction of Model, Numerical Example, and Further Comments. In addition, the application of each model is illustrated with at least one example — the data in most of these illustrative examples have been sanitized to maintain the confidentiality of the companies where the studies were originally undertaken.

This book is solidly based on the results of real-world research in physical asset management (PAM), including applications of the models presented in the text. The new knowledge thus created is firmly rooted in reality, and it appears for the first time in book form. Among the materials included in this book are models relating to spare-parts provisioning, condition-based maintenance, and replacement of equipment with varying levels of utilization. The risk of failure, characterized by the hazard function, is an important element in many of the models presented in this book. It is determined by fitting a suitable statistical model to life data. As Abernethy states, "Weibull analysis is the leading method in the world for fitting life data" (page 1-1, *The New Weibull Handbook*, second edition, Gulf Publishing Company, Houston, TX, 1996); Appendix 2 addresses Weibull analysis. This appendix contains a section that deals with trend analysis of life data, a vital issue to consider before undertaking a Weibull analysis.

To eliminate the tedium of performing the analysis manually, software that implements many of the procedures and models featured in this book has been developed. The educational versions of such software are packaged into MORE (Maintence, Optimization, and Reliability Engineering) tools that can be downloaded free from the publisher's Web site at http://www.crcpress.com/e_products/downloads/download.asp?cat_no=DK9669. These software packages include:

- OREST (acronym for Optimal Replacement of Equipment in the Short Term) — introduced in Chapter 2, Section 2.13
- SMS (acronym for Spares Management Software) — introduced in Chapter 2, Section 2.11.5

- PERDEC (acronym for Plant and Equipment Replacement Decisions) and AGE/CON (based on the French term *L'Age Économique*) — introduced in Chapter 4, Section 4.7
- Workshop Simulator — introduced in Chapter 5, Section 5.4.3
- Crew Size Optimizer — introduced in Chapter 5, Section 5.6.3
- WeibullSoft — introduced in Appendix 2, Section A2.6

This book can be used as a textbook for a one-semester senior undergraduate or postgraduate course on maintenance decision analysis. Problem sets with answers are provided at the end of each chapter that presents the decision tools. Additional resources are available to support the use of this book. These include an extensive set of PowerPoint slides covering the various chapters and Appendices 1 to 3, a solutions manual for the problems in the book, and a bank of more than 100 examination questions. Instructors who adopt the book can obtain these resources by contacting Susie Carlisle at susie.carlisle@taylorandfrancis.com.

If the book is used as a teaching text, many of the "Further Comments" sections should generate sufficient ideas for the reader to specify problems different from those given in the text, so that he can then practice the construction of mathematical models.

The book can also be used for a 3- to 4-day continuing professional development course for maintenance and reliability professionals. Such students may wish to omit the details on the formulation of the models and just focus on the "Applications" sections. They are advised to delve into the models only when they are prepared to invest the time and effort necessary to understand the underpinning theories — to borrow a statement articulated by an anonymous high school teacher, "Mathematical modeling is not a spectator sport."

The real-world applications given in Chapters 2 to 5 highlight the practical uses of the decision tools presented in this book. Readers interested in exploring the possibility of applying these tools or their extensions to address specific problems may find it useful to refer to the expanded list of applications given in Appendix 4.

With much data becoming available, we often find ourselves in the bewildering position of being data rich but information poor. We may have all the raw data we will ever need at our fingertips. However, unless we can interpret and use such data intelligently, it is of little use. To transform the data into information useful for decision making, we need the appropriate tools, such as those presented in this book.

The more you do, the more you can do. We suggest that maintenance and reliability professionals apply the knowledge acquired in this book initially to address a simple maintenance problem within their organization. In this manner, they can gain confidence in using the tools featured in this book, and later apply them in more challenging situations.

Andrew K.S. Jardine and Albert H.C. Tsang

Acknowledgments

We wish to acknowledge the financial support provided by the Hong Kong Polytechnic University and the Natural Sciences and Engineering Research Council of Canada (NSERC) that enabled the authors to meet from time to time, in both Canada and Hong Kong, as the book developed.

Readers may recognize some of the models that are featured in this book because they first appeared in *Maintenance, Replacement and Reliability*, a text of one of the authors. Through using that book in courses we taught, we maintained our keen interest in maintenance optimization. We gratefully acknowledge the support and insights of many current and former colleagues, research students, students and participants of our courses, with whom we have interacted in maintenance optimization discussions and teaching, in contributing to the creation of the new materials in this publication. Our gratitude also extends to those industries that supported projects we supervised or which sponsored participants to attend post-experience courses/seminars/workshops delivered by us in many countries, covering all continents (Africa, Asia, Australia, Europe, the Americas) except Antarctica. Applications cited in the book are, with one exception, based on studies we have undertaken. In all such cases, the data has been sanitized to maintain confidentiality of the source. We thank the late John D. Campbell for the consultancy opportunities afforded by his International Center of Excellence in Maintenance Management at PricewaterhouseCoopers, and subsequently IBM's Center of Excellence in Asset Management, some of which resulted in applications cited in the book.

The support of many colleagues has been instrumental to the successful completion of this book. In particular, we wish to acknowledge: Dragan Banjevic and Darko Louit who provided valuable support in the preparation of Section 2.11 (Spare Parts Provisioning: Insurance Spares); Baris Balcioglu, Dragan Banjevic, Renyan Jiang, Darko Louit, Diederik Lughtiheid, Daming Lin, Neil Montgomery, Emanuel Bernhard, and Ali Zuashkianu who reviewed various chapters of the book as it was being developed; Eric Wing Kin Lee, Jesus Tamez Navarro, and Wai Keung Tang who created the program codes of WeibullSoft, Workshop Simulator, and Crew Size Optimizer; Peter Lau for packaging the MORE (Maintenance Optimization & Reliability Engineering) Tools, the suite of software tools that can be downloaded from the Web by users of this book; Diane Gropp, assisted by Susan Gropp, who prepared all the text and drawings in final form to meet the requirements of the publisher; Samantha Chan who transcribed an early draft of the book; and Peter Jackson who did the copyediting to ensure consistency of style in the prose. We are also grateful to colleagues who suggested entries in the references and further reading lists in the book. Also, Andrey Pak had the unenviable task of carefully reviewing the complete text, and preparing the extensive Instructors' Manual. He did a fine job, and any errors that remain should be seen as ours.

Permission has been obtained to include the following copyrighted materials in the book:

Andrew K.S. Jardine and Albert H.C. Tsang

About the Authors

Andrew K.S. Jardine is Professor and Principal Investigator at the Condition-Based Maintenance (CBM) laboratory in the Department of Mechanical and Industrial Engineering at the University of Toronto where the EXAKT software for CBM optimization has been developed. Professor Jardine undertook his undergraduate engineering study at the University of Strathclyde, Scotland, and his PhD was awarded by the University of Birmingham, England. He is a registered professional engineer in Canada and a chartered engineer in the United Kingdom, being a member of both the Institution of Mechanical Engineers and the Institution of Electrical Engineers. Professor Jardine is also a senior member of the Institute of Industrial Engineers and a member of the Operational Research Society

Professor Jardine also serves as a Subject Matter Expert to the Asset Management Center of Excellence of IBM Business Consulting Services. He is the author of the economic life software AGE/CON and PERDEC that is licensed globally to organizations that include the transportation, mining, electrical utilities, and process industries. Additionally he authored the OREST software for component replacement decision making. In 2001 he co-edited with J.D. Campbell the book *Maintenance Excellence: Optimizing Equipment Life Cycle Decisions.*

In 1993 Professor Jardine was the Eminent Speaker to the Maintenance Engineering Society of Australia and in 1998 was the first recipient of the Sergio Guy Memorial Award from the Plant Engineering and Maintenance Association of Canada "in recognition of his outstanding contribution to the maintenance profession."

Albert H.C. Tsang is Principal Lecturer in the Department of Industrial & Systems Engineering at The Hong Kong Polytechnic University, and Leader of its BEng (Hons) in Industrial & Systems Engineering program. He has a PhD from the University of Toronto. Dr. Tsang is a chartered engineer in the United Kingdom with working experience in the manufacturing industry covering functions such as industrial engineering, quality assurance, and project management. He is a founding member, past Chairman, serving Executive Committee member, and Fellow of the Hong Kong Society for Quality (HKSQ). Dr. Tsang has provided consultancy and advisory services to organizations in public utilities, health care, and government sectors on matters related to quality, reliability, maintenance, performance management, and assessment of performance excellence.

Dr. Tsang is the author of "WeibullSoft," a computer-aided self-learning package on Weibull analysis, and *Reliability-Centred Maintenance: A Key to Maintenance Excellence*, a book published in 2000 with Andrew K.S. Jardine, J.D. Campbell, and J.V. Picknell as co-authors.

Abstract

Reliability-centered maintenance (RCM) determines the type of maintenance tactics to be applied to an asset for preserving system function. While it answers the question of "What type of maintenance action needs to be taken?" the issue of when to perform the recommended maintenance action that will produce the best results remains to be addressed. Taking a longer-term perspective, we have to make decisions on asset replacement in the best interests of the organization, and determine the resource requirements of the maintenance operation that will meet business needs in a cost-effective manner. This book shows how data-driven procedures and tools can be used to address these important optimization issues in the organization's pursuit of excellence in asset management.

A framework that organizes the key areas of maintenance and replacement decisions is presented in the beginning, setting the scene for the range of problems covered in the book. This is followed with discussions that highlight the principles associated with optimization, model construction, and analysis. The problem areas studied include preventive replacement intervals, inspection frequencies, condition-based maintenance actions, capital equipment replacement, and maintenance resource requirements. The models presented are firmly rooted in reality, as they are based on the results of real-world research. The relevant statistics, Weibull analysis tools, and time value of money concepts that support formulation of maintenance models are given in the appendices.

There is a developing demand in universities and colleges for courses in the general areas of reliability, maintainability, enterprise asset management, physical asset management, and reliability and maintainability engineering. This book will have a significant role to play in such courses. It will also meet the increasing demand of practicing maintenance and reliability professionals for knowledge of tools that can be used to optimize their maintenance and reliability decisions.

Contents

1 Introduction

The two good rules of modeling:
Clearly define the question to be answered with the model
Make the model no more complex than necessary to answer the question

John Harte

1.1 FROM MAINTENANCE MANAGEMENT TO PHYSICAL ASSET MANAGEMENT

According to the classical view, the role of maintenance is to fix broken items. Taking such a narrow perspective, maintenance activities will be confined to the reactive tasks of repair actions or item replacement triggered by failures. Thus, this approach is known as reactive maintenance, breakdown maintenance, or corrective maintenance. A more recent view of maintenance is defined by Geraerds (1985) as "all activities aimed at keeping an item in, or restoring it to, the physical state considered necessary for the fulfilment of its production function." Obviously, the scope of this enlarged view also includes the proactive tasks, such as routine servicing and periodic inspection, preventive replacement, and condition monitoring. Depending on the deployment of responsibilities within the organization, these maintenance tasks may be shared by several departments. For instance, in an organization practicing total productive maintenance (TPM) (Nakajima, 1988), the routine servicing and periodic inspection of equipment are handled by the operating personnel, whereas overhauls and major repairs are done by the maintenance department. TPM will be discussed in more detail in Section 1.4.

If the strategic dimension of maintenance is also taken into account, it should cover those decisions taken to shape the future maintenance requirements of the organization. Equipment replacement decisions and design modifications to enhance equipment reliability and maintainability are examples of these activities. The Maintenance Engineering Society of Australia (MESA) recognizes this broader perspective of maintenance and defines the function as "the engineering decisions and associated actions necessary and sufficient for the optimization of specified capability." *Capability* in the MESA definition is the ability to perform a specific action within a range of performance levels. The characteristics of capability include function, capacity, rate, quality, responsiveness, and degradation. The scope of maintenance management, therefore, should cover every stage in the life cycle of technical systems (plant, machinery, equipment, and facilities):

specification, acquisition, planning, operation, performance evaluation, improvement, and disposal (Murray et al., 1996). When perceived in this wider context, the maintenance function is also known as physical asset management (PAM).

1.2 THE CHALLENGES OF PHYSICAL ASSET MANAGEMENT

The business imperative for organizations seeking to achieve performance excellence demands that these organizations continuously enhance their capability to create value for customers and improve the cost-effectiveness of their operations. PAM, an important support function in businesses with significant investments in plants and machinery, plays an important role in meeting this tall order.

The performance demanded of physical asset management has become more challenging as a result of the three developments discussed below (Tsang et al., 2000).

1.2.1 EMERGING TRENDS OF OPERATION STRATEGIES

The conventional wisdom of embracing the concept of economy of scale is losing followers. An increasing number of organizations have switched to lean manufacturing, just-in-time production, and six-sigma programs. These trends highlight a shift of emphasis from volume to quick response, elimination of waste, and defect prevention. With the elimination of buffers in such demanding environments, breakdowns, speed loss, and erratic process yields will create immediate problems to the timely supply of products and services to customers. Installing the right equipment and facilities, optimizing the maintenance of these assets, and the effective deployment of staff to perform the maintenance activities are crucial factors to support these operation strategies.

1.2.2 TOUGHENING SOCIETAL EXPECTATIONS

There is widespread acceptance, especially in the developed countries, of the need to preserve essential services, protect the environment, and safeguard people's safety and health. As a result, a wide range of regulations have been enacted in these countries to control industrial pollution and prevent accidents in the workplace. Scrap, defects, and inefficient use of materials and energy are sources of pollution. They are often the result of operating plant and facilities under less than optimal conditions. Breakdowns of mission-critical equipment interrupt production. In chemical production processes, a common cause of pollution is the waste material produced during the start-up period after production interruptions. Apart from producing waste material, catastrophic failures of operating plants and machinery are also a major cause of outages of basic services, industrial accidents, and health hazards. Keeping facilities in optimal condition and preventing critical failures are an effective means to manage the risks of service interruptions, pollution, and industrial accidents. These are part of the core functions of PAM.

1.2.3 TECHNOLOGICAL CHANGES

Technology has always been a major driver of change in diverse fields. It has been changing at a breathtaking rate in recent decades, with no signs of slowing down in the foreseeable future. Maintenance is inevitably under the influence of rapid technological changes. Nondestructive testing, transducers, vibration measurement, thermography, ferrography, and spectroscopy make it possible to perform nonintrusive inspection. By applying these technologies, the condition of equipment can be monitored continuously or intermittently while it is in operation. This has given birth to condition-based maintenance, an alternative to the classical time-driven approach to preventive maintenance.

Power electronics, programmable logic controllers (PLCs), computer controls, transponders, and telecommunications systems are used to substitute electromechanical systems, producing the benefits of improved reliability, flexibility, compactness, light weight, and low cost. Fly-by-wire technology, utilizing software-controlled electronic systems, has become a design standard for the current generation of aircraft. Flexible manufacturing cells and computer-integrated manufacturing systems are gaining acceptance in the manufacturing industry. In some of the major cities, contactless smartcards (CSCs) have gained acceptance as a convenient means of making payments. In the electric utility industry, automation systems are available to remotely identify and deal with faults in the transmission and distribution network. Radio frequency identification (RFID) technology can be deployed to track mobile assets such as vehicles. Data transmitted to RFID tags from sensors embedded in mission-critical assets can be used for health monitoring and prognosis.

The deployment of these new technologies is instrumental to enhancing system availability, improving cost-effectiveness, and delivering better or innovative services to customers. The move presents new challenges to asset management. New knowledge has to be acquired to specify and design these new technology-enabled systems. New capability has to be developed to commission, operate, and maintain such new systems. During the phase-in period, interfacing old and new plants and equipment is another challenge to be handled by the physical asset management function.

1.3 IMPROVING PHYSICAL ASSET MANAGEMENT

To meet the challenges identified in Section 1.2, organizations need to focus on improving the performance of their physical assets. This can be accomplished by having a clear strategy, the right people and systems, appropriate tactics, and controlled work through planning and scheduling, maintenance optimization, and process reengineering.

1.3.1 MAINTENANCE EXCELLENCE

A survey conducted by *Plant Engineering & Maintenance Magazine* (Robertson and Jones, 2004) indicated that maintenance budgets ranged from 2 to 90% of the total plant operating budget, with the average being 20.8%. It can be reasoned that operations and maintenance (O&M) represent a major cost item in equipment-intensive industrial operations. These operations can achieve significant savings in O&M costs by making the right and opportune maintenance decisions. Maintenance is often the business process that has not been optimized. Instead of being a liability of business operations, achieving excellence in maintenance will pay huge dividends through reduced waste and maximized efficiency and productivity, thereby improving the bottom line. Maintenance excellence is many things, done well. It happens when:

- A plant performs up to its design standards and equipment operates smoothly when needed.
- Maintenance costs are within budget and investment in new assets is reasonable.
- Service levels are high.
- Turnover of maintenance, repair, and operation (MRO) materials inventory is fast.
- Tradespersons are motivated and competent.

Most important of all, maintenance excellence is concerned with balancing performance, risks, and the resource inputs to achieve an optimal solution. This is not an easy task because much of what happens in an industrial environment is characterized by uncertainties.

The structured approach to achieving maintenance excellence is shown in Figure 1.1 (Campbell, 1995). There are three types of goals on the route to maintenance excellence (Campbell and Jardine, 2001), and they are discussed in the sections below.

1.3.1.1 Strategic

First, you must draw a map and set a course for your destination. The map comprises a vision of the asset management performance to be achieved and an assessment of the current level of performance; the difference between the two is known as the performance gap. The asset management strategy embraced by the organization informs the course of action for closing the gap. The resource requirements and time frame also need to be considered in developing the action plans. These management activities provide leadership for the maintenance effort and are depicted as the first layer in Figure 1.1.

1.3.1.2 Tactical

With the assets in place to support operations, you need a work management (planning and scheduling) and materials management system to control maintenance processes. Tactics to manage the risk of asset failures are selected. The

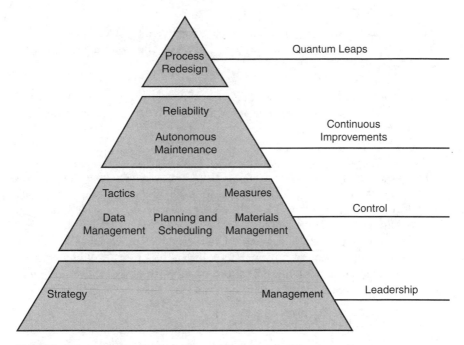

FIGURE 1.1 A structured approach to achieving maintenance excellence.

options include time-based maintenance actions, time-based discard, condition-based maintenance (CBM), run-to-failure, fault-finding tests, and process or equipment redesign. Data relating to equipment histories, warranty, and regulatory requirements, as well as status of maintenance work orders, must be documented and controlled. Typically, such data are managed by a computerized maintenance management system (CMMS) or enterprise asset management (EAM) system.

Performance indicators relating to various aspects of the maintenance service are tracked in order to evaluate performance of asset management (see, for example, Wireman, 1999). Ideally, the measurement system must be holistic, and apart from providing information for process control, it should also influence people's behavior so that their efforts are aligned with the strategic intent of the organization's asset management. The balanced scorecard (Kaplan and Norton, 1996) provides such an approach to measurement, and is developed on the notion that no single measure is sufficient to indicate the total performance of a system. It translates the organization's strategy on maintenance into operational measures in multiple dimensions (such as financial, safety, users, internal processes, and organizational development) that collectively are critical indicators of current achievements as well as powerful drivers and predictors of future asset performance. Examples of balanced scorecards for measuring asset management performance can be found in Niven (1999) and Tsang and Brown (1999).

1.3.1.3 Continuous Improvements

In pursuit of continuous improvement, two complementary methodologies that reflect different focuses are available to enhance reliability (uptime) of physical assets. These methodologies are:

- Total productive maintenance (TPM) — a people-centered methodology
- Reliability-centered maintenance (RCM) — an asset-centered methodology

They are discussed in Sections 1.4 and 1.5, respectively.

Decisions are to be made on when to perform the selected maintenance action and how much resource to be deployed to meet the expected maintenance demands. Instead of relying on intuition-based pronouncements, such as strength of personalities or the number of complaints received from mechanics, fact-based arguments should be used in making these maintenance decisions. Decisions driven by information extracted from data will lead to optimal solutions. Thus, data management, featured in level 2 of the structured approach shown in Figure 1.1, plays an important role in supporting decision optimization.

1.3.2 QUANTUM LEAPS

Finally, by engaging the collective wisdom and experience of the entire workforce, adopting the best practices that exist within and outside your organization, and redesigning the work processes, the organization will set in motion breakthrough changes that make quantum leaps in asset management performance. See, for example, Campbell (1995) for a detailed discussion of these efforts.

1.4 RELIABILITY THROUGH THE OPERATOR: TOTAL PRODUCTIVE MAINTENANCE

TPM is a people-centered methodology that has proven to be effective for optimizing equipment effectiveness and eliminating breakdowns. It mobilizes the machine operators to play an active role in maintenance work by cultivating in these frontline workers a sense of ownership of the facilities they operate (Campbell, 1995) and enlarging their job responsibilities to include routine servicing and minor repair of their machines. Through this type of operator participation in maintenance activities, TPM aims to eliminate the six big losses of equipment effectiveness, see Table 1.1 (Nakajima, 1988). In the manufacturing sector, 15 to 40% of total manufacturing costs are maintenance related. At least 30% of these costs can be eliminated through TPM.

To achieve zero breakdowns, hidden defects in the machine need to be exposed and corrected before they have deteriorated to the extent that they will cause the machine to break down. This can be accomplished by maintaining equipment in good basic condition through proper cleaning and effective lubrication,

TABLE 1.1
Six Big Losses of Equipment Effectiveness

Breakdowns
Setup and adjustment
Idling and minor stoppages
Reduced speed
Defects in process
Reduced yield

restoring the condition of deteriorated parts, and enhancing the operation, setup, inspection, and maintenance skills of operators. Traditionally, these duties fall outside the responsibilities of the machine operator, whose role is nothing else but to operate the machine; when it breaks down, his duty is to request maintenance to fix it. Thus, TPM involves a restructuring of work relating to equipment maintenance. Machine operators are empowered to perform routine inspection, servicing, and minor repairs. This concept of operator involvement in enhancing equipment wellness is known as autonomous maintenance (AM). It is cultivated through 5S and CLAIR.

5S is a tool for starting the journey toward world-class competitiveness. It is a team effort that involves everyone in the organization to create a productive workplace by keeping it safe, clean, and orderly. 5S stands for:

- Sorting
  - Separate the needed from the not needed.
  - Identify items that you use frequently. Sort, tag, and dispose of the unneeded items.
- Simplifying
  - A place for everything, and everything in its place.
  - Once you have determined what you need, organize it and standardize its use to increase your effectiveness.
- Systematic cleaning
  - Making things ready for inspection.
  - Regular cleaning helps to solve problems before they become too serious by identifying sources and root causes. Having a clean, well-organized workplace also makes work more efficient and more productive — whether on the production line or in customer service.
- Standardizing
  - Create common methods to achieve consistency.
- Sustaining
  - Constant maintenance, improvement, and communication.

5S becomes a continuous improvement process. Readers interested in 5S implementation may refer to Hirano (1990).

CLAIR is an acronym for clean, lubricate, adjust, inspect, minor repair. The concept is to have operators work with maintenance toward the common goals of stabilizing equipment conditions and halting accelerated deterioration. The operators are empowered to perform the basic tasks of cleaning, checking lubrication, simple adjustments, inspections and replacement of parts, minor repairs, and other simple maintenance tasks. By providing them with training on equipment functions and functional failures, the operators will also prevent failure through early detection and treatment of abnormal conditions.

Turning operators into active partners with maintenance and engineering to improve overall performance and reliability of equipment is a revolutionary concept. Thus, training, slogans, and other promotional media — activity boards, one-point lessons, photos, cartoons — are typically used to create and sustain the cultural change.

Being relieved of the routine tasks of maintenance, the experts in the maintenance unit can be deployed to focus on more specialized work, such as major repairs, overhauls, tracking and improving equipment performance, and replacement or acquisition of physical assets. Instead of having to continuously fight fires and attend to numerous minor chores, the unit can now devote its resources to addressing strategic issues such as formulation of maintenance strategies, establishment of maintenance management information systems, tracking and introduction of new maintenance technologies, and training and development of production and maintenance workers.

A full discussion of TPM is outside the scope of this book. Readers interested in the topic can refer to Dillon (1997), Nakajima (1988), Tajiri and Gotoh (1992), and Tsang and Chan (2000).

1.5 RELIABILITY BY DESIGN: RELIABILITY-CENTERED MAINTENANCE

TPM has a strong focus on people and the basics, such as cleaning, tightening, and lubricating, for ensuring the well-being of equipment. Its emphasis is on early detection of wear-out to prevent in-service failures. RCM is an alternative approach to enhancing asset reliability by focusing on design. It asks questions such as: Do we have to do maintenance at all? Will a design change eliminate the root cause of failure? What kind of maintenance is most likely to meet the organization's business objectives?

RCM is a structured methodology for determining the maintenance requirement of a physical asset in its operating context. The asset can be part of a larger system. The primary objective of RCM is to preserve system function rather than to keep an asset in service. Application of RCM requires a full understanding of the functions of physical assets, and the nature of failures related to these functions. It recognizes that not all failures are created equal, and some failures cannot be prevented by overhaul or preventive replacement. Thus, maintenance actions that are not cost-effective in preserving system function will not be performed.

RCM can produce these benefits:

- Improve understanding of the equipment — how it fails and the consequences of failure
- Clarify the roles that operators and maintainers play in making equipment more reliable and less costly to operate
- Make the equipment safer, more environmentally friendly, more productive, more maintainable, and more economical to operate

The following results of RCM applications have been reported in various industry sectors (Tsang et al., 2000):

- Manufacturing
  - Reduced routine preventive maintenance requirements by 50% at a confectionery plant
  - Increased availability of beer packaging line by 10% in 1 year
- Utility
  - Reduced maintenance costs by 30 to 40%
  - Increased capacity by 2%
  - Reduced routine maintenance by 50% on 11-kV transformers
- Mining
  - Reduced annual oil filter replacement costs in haul truck fleet by US$150K
  - Reduced haul truck breakdowns by 50%
- Military
  - Ship availability increased from 60 to 70%
  - Reduced ship maintenance requirements by 50%

The RCM methodology develops the appropriate maintenance tactics through a thorough and rigorous decision process, as shown in Figure 1.2.

Step 1: Select and prioritize equipment
Production and supporting processes are examined in order to identify key physical assets. These key physical assets are then prioritized according to how critical they are to operations, cost of downtime, and cost to repair.

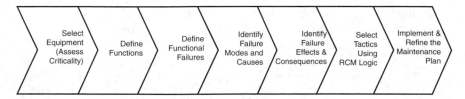

FIGURE 1.2 The RCM process.

Step 2: Define functions and performance standards
The functions of each system selected for RCM analysis need to be defined. The functions of equipment are what it does. It is important to note that some systems are dormant until some other event occurs, as in safety systems. Each function also has a set of operating limits. These parameters define normal operation of the function under a specified operating environment.

Step 3: Define functional failures
When the system operates outside its normal parameters, it is considered to have failed. Defining functional failures follows from these limits. We can experience our systems failing when they are high, low, on, off, open, closed, breached, drifting, unsteady, stuck, and so forth. Furthermore, failures can be total, partial, or intermittent.

Step 4: Identify failure modes/root causes
A failure mode is how the system fails to perform its function. A cylinder may be stuck in one position because of a lack of lubrication by the hydraulic fluid in use. The functional failure in this case is the failure to provide linear motion, but the failure mode is the loss of lubricant properties of the hydraulic fluid. It should be noted that a failure may have more than one possible root cause. This step identifies the chain of events that happen when a failure occurs. These questions are relevant in the analysis: What conditions needed to exist? What event was necessary to trigger the failure?

Step 5: Determine failure effects and consequences
This step determines what will happen when a functional failure occurs. The severity of the failure's impact on safety, the environment, operation, and maintenance is assessed.

The results of analyses made in steps 2 to 5 are documented in a failure mode, effect, and criticality analysis (FMECA)* worksheet (Stamatis, 2003).

Step 6: Select maintenance tactics
Maintenance actions are performed to mitigate functional failures. A decision logic tree is used to select the appropriate maintenance tactics for the various functional failures. Before finalizing the tactic decision, the other technically feasible alternatives need to be considered in order to determine the one that is most economical. Figure 1.3 summarizes the RCM logic. If time-based maintenance intervention or periodic inspection has been selected, the frequency of such task needs to be determined in order to

---

* Apart from FMECA, there are other methodologies for assessing and managing risks relating to operation and maintenance of physical assets. These include hazard and operability studies (HAZOPS) (Kletz, 1999), fault tree analysis (FTA) (CAN/CSA-Q636-93, 1993), and, in the case of the petro-chemical industrial, risk-based inspection (RBI) (ASME, 2003).

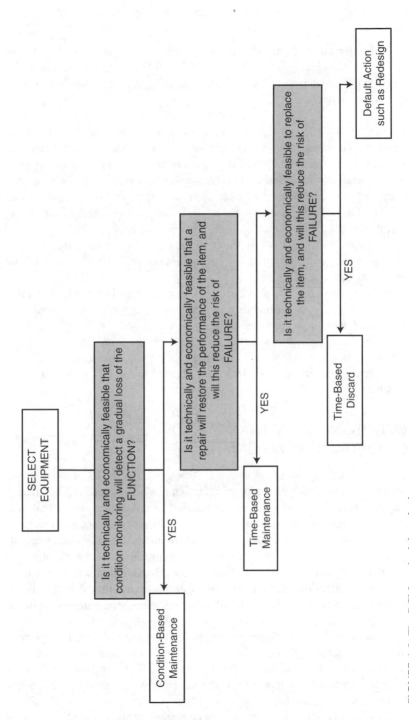

FIGURE 1.3 The RCM methodology logic.

achieve optimal results. This will be discussed in the subsequent chapters of this book.

Step 7: Implement and refine the maintenance plan
The maintenance plan developed in step 6 is implemented and the results are reviewed to determine if the plan needs to be refined or modified to ensure its effectiveness.

Implementation of RCM requires the formation of a multidisciplinary team with members knowledgeable in the day-to-day operations of the plant and equipment, as well as in the details of the equipment itself. This demands at least one operator and one maintainer. Members with knowledge of planning and scheduling and overall maintenance operations and capabilities are also needed to ensure that the tasks are truly doable in the plant environment. Thus, senior-level operations and maintenance representation is also needed. Finally, detailed equipment design knowledge is important to the team. This knowledge requirement generates the need for an engineer or senior technician/technologist from maintenance or production.

Before the analysis begins, the RCM team should determine the plant baseline measures for reliability and availability, as well as the coverage and compliance of a proactive maintenance program. These measures will be used later when comparing what has been changed and the success it is achieving.

Further discussion of RCM is beyond the scope of this book. Readers interested in the topic can refer to SAE JA1011 (1999), Moubray (1997), and Smith and Hinchliffe (2004).

1.6 OPTIMIZING MAINTENANCE AND REPLACEMENT DECISIONS

RCM determines the type of maintenance tactics to be applied to an asset. While it answers the question of "What type of maintenance action needs to be taken?" the issue of when to perform the recommended maintenance action that will produce the best results possible remains to be addressed. Taking a longer-term perspective, we have to make decisions on asset replacement in the best interests of the organization and determine the resource requirements of the maintenance operation that will meet business needs in a cost-effective manner. The optimization of these tactical decisions is the important issue addressed in the top of the "continuous improvements" layer of the maintenance excellence pyramid shown in Figure 1.1.

Traditionally, maintenance practitioners in industry are expected to cope with maintenance problems without seeking to operate in an optimal manner. For example, many preventive maintenance schemes are put into operation with only a slight, if any, quantitative approach to the scheme. As a consequence, no one is very sure just what the best frequency of inspection is or what should be inspected, and as a result, these schemes are cancelled because it is said they cost too much. Clearly some form of balance between the frequency of inspection and the returns from it is required (for example, fewer breakdowns since minor faults are detected before they result in costly repairs). In the subsequent chapters of this book we will examine various maintenance problem areas, noting the conflicts that ought to be considered and illustrating how they can be resolved in a quantitative manner in order to achieve optimal or near-optimal solutions to the problems. Thus, we indicate ways in which maintenance decisions can be optimized, where optimization is defined as attempting to resolve the conflicts of a decision situation in such a way that the variables under the control of the decision maker take the best possible values. Since the qualifier *best* is used, it is necessary to define its meaning in the context of maintenance. This will be covered in Section 1.7 of this chapter.

Asset managers who wish to optimize the life cycle value of the organization's human and physical assets must consider four key decision areas, which are shown as columns in Figure 1.4. The first column deals with component replacement, the second with inspection activities, including condition monitoring, and the third with replacement of capital equipment. The last column covers decisions concerning resources required for maintenance and their location.

Figure 1.4 forms the framework for Chapters 2 to 5 of this book. These chapters are devoted to the construction of mathematical models that are appropriate for different problem situations. The purpose of these mathematical models is to enable the consequences of alternative maintenance decisions to be evaluated

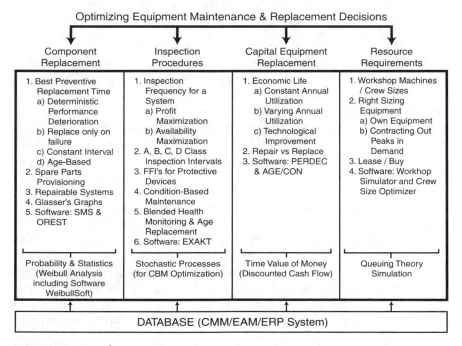

FIGURE 1.4 Key areas of maintenance and replacement decisions.

fairly rapidly in order to determine optimal decisions in relation to an objective. The problem areas covered are as follows:

Chapter 2: Component Replacement Decisions
This chapter covers determination of replacement intervals for equipment, the operating costs of which increase with use; the interval between preventive replacements of items subject to breakdown (also known as the group or block policy); and the preventive replacement age of items subject to breakdown.

Chapter 3: Inspection Decisions
This chapter covers determination of inspection frequencies for complex equipment used continuously; fault-finding intervals for protective devices; and condition-based maintenance (CBM) decisions.

Chapter 4: Capital Equipment Replacement Decisions
This chapter is concerned with determining the replacement intervals for capital equipment, the utilization pattern of which is fixed; replacement intervals for capital equipment, the utilization pattern of which is variable; and replacement policy for capital equipment taking into account technological improvement.

Chapter 5: Maintenance Resource Requirements
This chapter discusses problems relating to the determination of the mix of equipment to be installed in a maintenance workshop; the right size and composition of a maintenance crew; the extent of use of subcontracting opportunities; and lease or buy decisions.

In these chapters, we use a common format to present each decision optimization model. First, we provide a statement of the decision problem. A model that represents the essence of the decision problem is then presented. This is followed with a numerical example to illustrate the use of this model. In order to avoid unnecessary complications in developing the analytical model, various assumptions are made that, in practice, may not be applicable in some situations. Since the assumptions used in constructing the model are clearly stated, it is hoped that the reader will then be able to extend the simple model to fit specific problems. In this regard, we provide comments on further extension of the model to represent more details of the reality, if deemed necessary. For each model, with a few exceptions, we also provide one or more real-world application examples.

Some of these decision optimization models are made available as software programs that can be downloaded from the Web at http://www.crcpress.com/e_products/downloads/download.asp?cat_no=DK9669. In such cases, use of the software program in decision analysis is explained with at least one illustrative example.

Appendices 1 to 3 are included in the book to give a brief introduction to certain basic concepts and tools that must be understood before we can proceed to determine optimal maintenance procedures. Since uncertainty abounds in the area of maintenance (e.g., uncertainty about when equipment will fail), a knowledge of statistics and probability is required. An introduction to relevant statistics is given in Appendix 1. Modeling the risk of failure is a crucial step in optimizing replacement of components that are subject to failure. Weibull analysis, a powerful tool for modeling such risks, is introduced in Appendix 2. Appendix 3 deals with the present value concept. When making replacement decisions for capital equipment we take account of the fact that the value of a sum of money to be spent or received in the future is less than that if it is spent or received now. The present value concept is used to cover this fact. While within Chapters 2 to 5, applications of the tools featured in each chapter are highlighted, an expanded list of such applications is provided in Appendix 4. It serves to illustrate the breadth of actual applications that use the models or procedures presented in the book, or their extensions.

Optimizing maintenance and replacement decisions needs good quality and timely data. This need is depicted as the foundation of the framework shown in Figure 1.4. Such data are typically maintained in the database of the CMMS, EAM, or ERP. Readers interested in discussions of CMMS, EAM, and ERP in the context of physical asset management are referred to articles published on Web sites such as www.plant-maintenance.com and www.reliabilityweb.com.

1.7 THE QUANTITATIVE APPROACH

The primary purpose of using any quantitative discipline, such as industrial engineering, operational research, or systems analysis, is to assist management in decision making by using known facts more effectively, by enlarging the proportion of factual knowledge, and by reducing reliance on subjective judgment.

In the context of maintenance decision making, it is often found that very little factual knowledge is available. Although abundant data may have been captured in the organization's CMMS, EAM, or ERP, asset managers may not know the data-mining technique to extract useful knowledge from such data. This type of information is absolutely necessary for the development of optimal maintenance procedures. Appendix 2 introduces one such data-mining technique; it turns failure data into knowledge of the risk of failure of various assets.

There is keen interest in evidence-based maintenance decisions rather than the use of gut feeling or indiscriminately following the manufacturer's recommendations. It is hoped that this book will go some way toward reducing the proportion of subjective judgment in maintenance decision making.

As an early example of quantitative decision making in maintenance, which highlights the importance of selecting the correct objectives, we refer to a study undertaken during the Second World War by an operational research group of the Royal Air Force (see Crowther and Whiddington, 1963).

The specific problem was that performance of maintenance on Coastal Command aircraft was measured in terms of serviceability, the target of which was 75%. Serviceability was the ratio of the number of aircraft on the ground available to fly plus those flying, to the total number of aircraft. While a 75% serviceability was considered highly desirable, Coastal Command was also asked to get more flying time from aircraft. The Coastal Command Operational Research Section was called in to examine the problem. The section examined one cycle of operation of an aircraft and established that the aircraft could be in one of three possible states:

- Flying
- In maintenance
- Available to fly

Serviceability, S, which was the criterion of maintenance performance, was then

$$S = \frac{F + A}{F + A + M}$$

where F, A, and M are the average times that an aircraft spent in the flying, available to fly, and maintenance states, respectively. Further examination of the problem revealed that for every hour spent flying, 2 hours were required for maintenance. Using this information, it is possible to determine that to achieve a target of 75% serviceability, only 12.5% of an aircraft's time is spent flying, with 25% being spent on maintenance and 62.5% in an available state. However, if the serviceability is reduced to one third, then one third of the aircraft's time is spent flying, with two thirds of its time being spent in maintenance and 0% in the available state.

Thus, simply by aiming for a serviceability of one third, the flying hours could be considerably increased. Clearly in the scenario in which the Coastal Command aircraft were operating, the accepted objective of maintenance, namely, a high serviceability, was wrong. However, for other scenarios, such as the case in which aircraft are called on only in emergencies, a high serviceability objective may well be relevant.

As a result of the above analysis, instructions were given that, whenever possible, aircraft should be in the flying state, thus more than doubling the amount of flying time when compared to the mode of operation prior to examining the appropriateness of the maintenance objective.

1.7.1 Setting Objectives

One of the first steps in the use of quantitative techniques in maintenance is to determine the objective of the study. Once the objective is determined, whether it be to maximize profit/unit time, minimize downtime/unit time, maximize availability of protective devices subject to budgetary constraints, and so on, an evaluative mathematical model can be constructed that enables management to determine the best way to operate the system to achieve the required objective.

In the planned flying–planned maintenance study referred to above, Coastal Command's maintenance objective originally was to achieve a serviceability of 75%, but the study made it clear that this was the wrong objective, and what they should have been aiming for was a serviceability of one third to achieve more flying hours.

Also, in the study it was mentioned that a high serviceability was perhaps relevant to aircraft called on only in an emergency. This stresses the point that the objective a system is operated to achieve may change through changes in circumstances. In the context of maintenance procedures, the way in which equipment is maintained when it is already operating at full capacity may well be different from the way it should be maintained under conditions of an economic slump.

In Chapters 2 to 5, the models of various maintenance problems are constructed in such a way that the maintenance procedures that are geared to enable profits to be maximized, total maintenance cost to be minimized, and so forth, can be identified. However, it must be emphasized that when determining optimal maintenance procedures, care must be taken to ensure that the objective being pursued is appropriate. For example, it will not be suitable for the asset management department to pursue a policy designed to minimize downtime of equipment if the organization requires a policy designed to maximize profit (as in the midst of an economic slump). The two policies may in fact be identical, but this is not necessarily so. This point will be demonstrated by an example in Chapter 3, Section 3.3.

1.7.2 MODELS

One of the main tools in the scientific approach to management decision making is that of building an evaluative model, usually mathematical, whereby a variety of alternative decisions can be assessed. Any model is simply a representation of the system under study. In the application of quantitative techniques to management problems, the type of model used is frequently a symbolic model where the components of the system are represented by symbols and the relationships of these components are described by mathematical equations.

To illustrate this model-building approach, we will examine a maintenance stores problem that, although simplified, will illustrate two of the most important aspects of the use of models: the construction of a model of the problem being studied, and its solution.

A STORES PROBLEM

A stores controller wishes to know how many items to order each time the stock level of an item reaches zero. The system is illustrated in Figure 1.5.

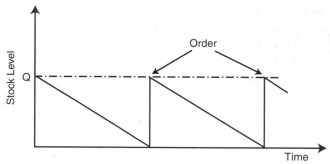

FIGURE 1.5 A stores problem.

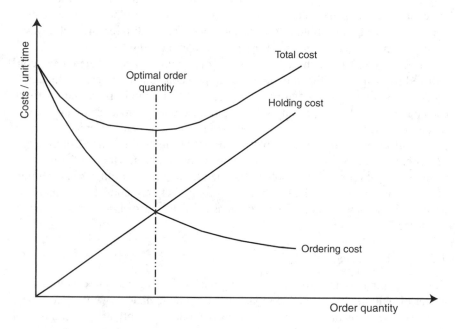

FIGURE 1.6 Optimal order quantity.

The conflicts in this problem are that the more items he orders at any time, the more his ordering costs will decrease, since he has to place fewer orders, but his stockholding costs will increase. These conflicting costs are illustrated in Figure 1.6.

The stores controller wants to determine the order quantity that minimizes the total cost. This total cost can be plotted, as shown in Figure 1.6, and used to solve the problem. In this particular case, the total cost is minimized when the order quantity is at the intersection of the holding cost curve and the ordering cost curve. However, this should not be generalized; for example, see Figure 1.8. A much more rapid solution to the problem, however, may be obtained by constructing a mathematical model. The following parameters can be defined:

D Total annual demand
Q Order quantity
C_o Ordering cost per order
C_h Stockholding cost per item per year

Optimal Order Quantity

Total cost per year of
  ordering and holding stock = Ordering cost per year + stockholding cost per year

Since

Ordering cost/year   = Number of orders placed per year × ordering cost per order

$$= \frac{DC_o}{Q}$$

Stockholding cost/year   = Average number of items in stock per year (assuming linear
                              decrease of stock) × stockholding cost per item per year

$$= \frac{1}{2} QC_h$$

Therefore, the total cost per year, which is a function of the order quantity, and denoted $C(Q)$, is

$$C(Q) = \frac{DC_o}{Q} + \frac{QC_h}{2} \qquad (1.1)$$

Equation 1.1 is a mathematical model of the problem relating order quantity Q to total cost $C(Q)$.

The stores controller wants the number of items to order in order to minimize the total cost, that is, to minimize the right-hand side of Equation 1.1. The answer comes by differentiating the equation with respect to Q, the order quantity, and equating the derivative to zero as follows:

$$\frac{dC(Q)}{dQ} = -\frac{DC_o}{Q^2} + \frac{C_h}{2} = 0$$

Therefore,

$$\frac{DC_o}{Q^2} = \frac{C_h}{2}$$

$$Q = \sqrt{\frac{2DC_o}{C_h}} \qquad\qquad (1.2)$$

Since the values of D, C_o, and C_h are known, substitution of them into Equation 1.2 gives the optimal value of Q. Strictly speaking, we should check that the value of Q obtained from Equation 1.2 is a minimum and not a maximum. The interested reader can check that this is the case by taking the second derivative of $C(Q)$ and noting that the result is positive. In fact, in this particular case, the optimal order quantity equalizes the average holding and ordering costs.

Example

Let $D = 1,000$ items, $C_o = \$5.00$, and $C_h = \$0.25$:

$$Q = \sqrt{\frac{2 \times 1000 \times 5}{0.25}} = 200 \text{ items}$$

Thus, each time the stock level reaches zero, the stores controller should order 200 items to minimize the total cost per year of ordering and holding stock.

Note that various assumptions have been made in the inventory model presented, which in practice may not be realistic. For example, no consideration has been given to the possibility of quantity discounts, the possible lead time between placing an order and its receipt, the fact that demand may not be linear, and the fact that demand may not be known with certainty. The purpose of the above model is simply to illustrate the construction and solution of a model for a particular problem. If the reader is interested in the stock control aspects of maintenance stores, see Nahmias (1997).

1.7.3 Obtaining Solutions from Models

In the stores problem of the previous section, two methods for solving a mathematical model were demonstrated, an analytical and a numerical procedure.

The calculus solution was an illustration of an analytical technique where no particular set of values of the control variable (amount of stock to order) was considered, but we proceeded straight to the solution given by Equation 1.2.

In the numerical procedure, solutions for various values of the control variables are evaluated in order to identify the best result; i.e., it is a trial-and-error procedure. The graphical solution of Figure 1.6 is equivalent to inserting different values of Q into the model (Equation 1.1) and plotting the total cost curve to identify the optimal value of Q.

In general, analytical procedures are preferred to numerical ones, but because of problem complexity, in many cases they are impracticable or even impossible to use. In many of the maintenance problems examined in this book, the solution to the mathematical model will be obtained by using numerical procedures. These are primarily graphical procedures, but iterative procedures and simulation are also used.

Perhaps one of the main advantages of graphical solutions is that they often enable management to see clearly the effect of implementing a maintenance policy that deviates from the optimum identified through solving the model. Also, it may be possible to plot the effects of different maintenance policies together, thus illustrating the relative impacts of the policies. To illustrate this point, Chapter 2 includes the analyses of two different replacement procedures:

1. Replacement of items at fixed intervals of time
2. Replacement of items based on the length of time they are actually in use

Intuitively, one might feel that procedure 2 would be preferable since it is based on usage of the item (thus preventing an almost new item from being replaced shortly after its installation subsequent to a previous failure, as would happen with procedure 1).

For these different maintenance policies, which can be adopted for the same equipment, models can be constructed, as is done in Chapter 2, and for each policy the optimal procedure can be determined. However, by using a graphical solution procedure, the maintenance cost of each policy can be plotted, as illustrated in Figure 1.7, and the maintenance manager can see exactly the effect of the alternative policies on total cost. It may well be the case that from a data collection point of view, one policy involves considerably less work than the other, yet they may have almost the same minimum total cost. This is illustrated in Figure 1.7, where the minimum total costs are about the same for procedures 1 and 2.

Of course, for different costs, breakdown distributions, failure and preventive replacement times, and so on, the minimum total costs and replacement intervals may differ greatly between different replacement policies. The point is that a

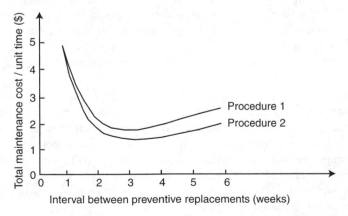

FIGURE 1.7 Comparing the total maintenance costs of two preventive replacement procedures.

graphical illustration of the solutions often assists the manager to determine the policy to be adopted. Also, such a method of presenting a solution is often more acceptable than a statement such as "policy x is the best," which may be presented along with complicated mathematics.

Further comments about the benefits of curve plotting are given in Section 2.2.4 of Chapter 2 in relation to the problem of determining the optimal replacement interval for equipment, the operating cost of which increases with use.

One of the developments in numerical procedures made possible by computers is simulation. An application of this procedure will be illustrated in a problem of Chapter 5 that relates to determining the optimal number of machines to be installed in a workshop.

1.7.4 MAINTENANCE CONTROL AND MATHEMATICAL MODELS

The primary function of maintenance is to control the condition of assets. Some of the problems associated with this include determination of:

Inspection frequencies
Overhaul intervals, i.e., part of a preventive maintenance policy
Whether to do repairs, i.e., a breakdown maintenance policy or not
Replacement rules for components
Replacement rules for capital equipment — perhaps taking account of technological change
Whether equipment should be modified
The size of the maintenance crew
Composition of machines in a workshop
Rules for the provision of spares

Appendix 4 provides a list of real-world applications of maintenance decision optimization models in different industries.

Problems within these areas can be classified as being deterministic or probabilistic. Deterministic ones are those where the consequences of a maintenance action are assumed to be nonrandom. For example, after an overhaul, the future trend in operating costs is known. A probabilistic problem is one where the outcome of the maintenance action is random. For example, after equipment repair, the time to next failure is uncertain.

To solve any of the above problems, there are often a large number of alternative decisions that can be taken. For example, for an item subject to sudden failure, we may have to decide whether to replace it while it is in an operating state, or only on its failure, whether to replace similar components in groups when perhaps one only has failed, and so on. Thus, the function of the asset management department is, to a large extent, concerned with determining the impact of various decisions to control the condition of assets, on meeting the objectives of the organization.

As indicated above, many control actions are open to the maintenance manager. The effect of these actions should not be looked at solely from their effect on the asset management department since the consequences of such actions may seriously affect other units of the organization, such as production or operations.

To illustrate the possible interactions of the asset management function on other departments, consider the effect of the decision to perform repairs only and not to do any preventive maintenance, such as overhauls. This decision may well reduce the budget for asset management, but it may also cause considerable production or operation downtime. In order to take account of interactions, sophisticated techniques are frequently required, and this is where the use of mathematical models can assist the maintenance manager and reduce tension that often occurs between maintenance and operations.

Figure 1.8 illustrates the type of approach taken by using a mathematical model to determine the optimal frequency of overhauling a piece of plant by balancing the input (maintenance cost) of the maintenance policy against its output (reduction in downtime).

The above example is very simple, and in practice, we have to consider many factors in the context of even a single maintenance decision. For example, if the objective of a maintenance decision is to minimize total costs — lowest cost optimization — the costs of the component or asset, labor, lost production, and perhaps even customer dissatisfaction from delayed deliveries are all to be considered. Where equipment or component wear-out is a factor, the lowest possible cost is usually achieved by replacing machine parts late enough to get good service out of them, but early enough for an acceptable rate of on-the-job failures (to attain a zero rate, one would probably have to replace parts every day). In another scenario where availability is to be maximized, one has to get the right balance between taking equipment out of service for preventive maintenance and suffering outages due to breakdowns. If safety is the most important factor, one might optimize for the safest

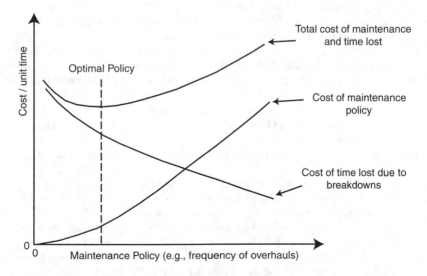

FIGURE 1.8 Optimal frequency of overhauls.

possible solution, but with an acceptable impact on cost. If profit is to be optimized, one would take into account not only cost, but the effect on revenues through greater customer satisfaction (better profits) or delayed deliveries (lower profits).

The example shown in Figure 1.8 should suffice to show that the quantitative approach taken in this book is concerned with determining appropriate maintenance decisions by studying the mathematical and statistical relationships between the decisions to be taken and the consequences of these decisions. The foregoing comments about the use of models for analyzing maintenance problems are very brief, but they will be elaborated further in the subsequent chapters of this book.

REFERENCES

ASME. (2003). *Risk-Based Methods for Equipment Life Management*. New York: American Society of Mechanical Engineers International.

Campbell, J.D. (1995). *Uptime: Strategies for Excellence in Maintenance Management*. Portland, OR: Productivity Press.

Campbell, J.D. and Jardine, A.K.S., Eds. (2001). *Maintenance Excellence: Optimizing Equipment Life Cycle Decisions*. New York: Marcel Dekker.

CAN/CSA-Q636-93. (1993). *Quality Management: Guidelines and Requirements for Reliability Analysis Methods*. Toronto: Canadian Standards Association.

Crowther, J.G. and Whiddington, R. (1963). *Science at War*. London: H.M.S.O.

Dillon, A.P. (1997). *Autonomous Maintenance for Operators*. Portland, OR: Productivity Press.

Geraerds, W.M.J. (1985). The cost of downtime for maintenance: preliminary considerations. *Maintenance Management International*, 5, 13–21.

Hirano, H. (1990). *5 Pillars of the Visual Workplace: The Source Book for 5S Implementation*. Portland, OR: Productivity Press.

Kaplan, R.S. and Norton, D.P. (1996). *The Balanced Scorecard: Translating Strategy into Action*. Boston, MA: Harvard Business School.

Kletz, T.A. (1999). *Hazop and Hazan: Identifying and Assessing Process Industry Hazards*, 4th ed. Rugby, England: Institute of Chemical Engineers.

Moubray, J. (1997). *Reliability Centred Maintenance*, 2nd ed. Oxford; Boston: Butterworth-Heinemann.

Murray, M., Fletcher, K., Kennedy, J., Kohler, P., Chambers, J., and Ledwidge, T. (1996). Capability Assurance: A Generic Model of Maintenance, ICOMS-96, Melbourne, Paper 72, pp. 1–5.

Nahmias, S. (1997). *Production and Operations Analysis*, 3rd ed. Chicago: Irwin/McGraw-Hill.

Nakajima, S. (1988). *Introduction to TPM*. Cambridge, MA: Productivity Press.

Niven, P.R. (1999). Cascading the balanced scorecard: a case study on Nova Scotia Power, Inc. *Journal of Strategic Performance Measurement*, 3(2), 5–12.

Robertson, R. and Jones, A. (2004). Pay day. *Plant Engineering & Maintenance*, 28(9), 18–25.

SAE JA1011. (1999). *Evaluation Criteria for Reliability-Centered Maintenance (RCM) Processes*. Warrendale, PA: Society of Automotive Engineers.

Smith, A.M. and Hinchliffe, G.R. (2004). *RCM: Gateway to World Class Performance.* Boston: Elsevier Butterworth-Heinemann.

Stamatis, D.H. (2003). *Failure Mode and Effect Analysis: FMEA from Theory to Execution*, 2nd ed. Milwaukee, WI: ASQ Quality Press.

Tajiri, M. and Gotoh, F. (1992). *TPM Implementation: A Japanese Approach.* New York: McGraw-Hill.

Tsang, A.H.C. and Brown, W.L. (1999). Managing the maintenance performance of an electricity utility through the use of balanced scorecards. *New Engineering Journal*, 2(3), 22–29.

Tsang, A.H.C. and Chan, P.K. (2000). TPM implementation in China: a case study. *International Journal of Quality & Reliability Management*, 17(2), 144–157.

Tsang, A.H.C., Jardine, A.K.S., Campbell, J.D., and Picknell, J.V. (2000). *Reliability Centred Maintenance: A Key to Maintenance Excellence.* Hong Kong: City University of Hong Kong.

Wireman, T. (1999). *Developing Performance Indicators for Managing Maintenance*, 1st ed. New York: Industrial Press.

2 Component Replacement Decisions

The squeaking wheel doesn't always get the grease. Sometimes it gets replaced.

Vic Gold

2.1 INTRODUCTION

The goal of this chapter is to present models that can be used to optimize component replacement decisions. The interest in this decision area is because a common approach to improving the reliability of a system, or complex equipment, is through preventive replacement of critical components within the system. Thus, it is necessary to be able to identify which components should be considered for preventive replacement, and which should be left to run until they fail. If the component is a candidate for preventive replacement, then the subsequent question to be answered is: What is the best time? The primary goal addressed in this chapter is that of making a system more reliable through preventive replacement. In the context of the framework of the decision areas addressed in this book, we are addressing column 1 of the framework, as highlighted in Figure 2.1.

Replacement problems (and maintenance problems in general) can be classed as either deterministic or probabilistic (stochastic).

Deterministic problems are those in which the timing and outcome of the replacement action are assumed to be known with certainty. For example, we may have an item that is not subject to failure but whose operating cost increases with use. To reduce this operating cost, a replacement can be performed. After the replacement, the trend in operation cost is known. This deterministic trend in costs is illustrated in Figure 2.2.

Examples of component replacement problems that can be treated with a deterministic model are provided in Table 2.1.

Probabilistic problems are those where the timing and outcome of the replacement action depend on chance. In the simplest situation, the equipment may be described as being *good* or *failed*. The probability law describing changes from good to failed may be described by the distribution of time between completion of the replacement action and failure. As described in Appendix 1, the time to failure is a random variable whose distribution may be termed the equipment's failure distribution.

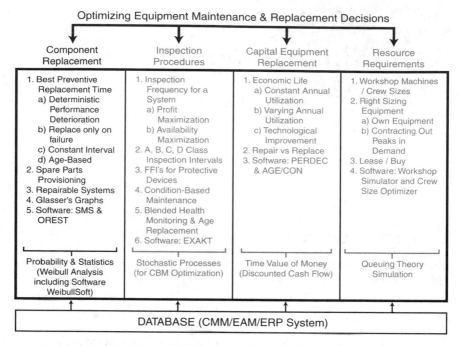

Component Replacement	Inspection Procedures	Capital Equipment Replacement	Resource Requirements
1. Best Preventive Replacement Time a) Deterministic Performance Deterioration b) Replace only on failure c) Constant Interval d) Age-Based 2. Spare Parts Provisioning 3. Repairable Systems 4. Glasser's Graphs 5. Software: SMS & OREST	1. Inspection Frequency for a System a) Profit Maximization b) Availability Maximization 2. A, B, C, D Class Inspection Intervals 3. FFI's for Protective Devices 4. Condition-Based Maintenance 5. Blended Health Monitoring & Age Replacement 6. Software: EXAKT	1. Economic Life a) Constant Annual Utilization b) Varying Annual Utilization c) Technological Improvement 2. Repair vs Replace 3. Software: PERDEC & AGE/CON	1. Workshop Machines / Crew Sizes 2. Right Sizing Equipment a) Own Equipment b) Contracting Out Peaks in Demand 3. Lease / Buy 4. Software: Workshop Simulator and Crew Size Optimizer
Probability & Statistics (Weibull Analysis including Software WeibullSoft)	Stochastic Processes (for CBM Optimization)	Time Value of Money (Discounted Cash Flow)	Queuing Theory Simulation

DATABASE (CMM/EAM/ERP System)

FIGURE 2.1 Component replacement decisions.

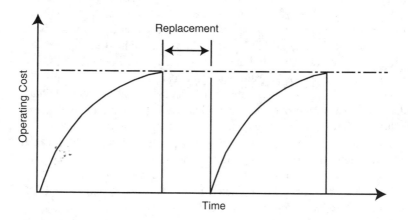

FIGURE 2.2 Deterministic trend in costs.

Examples of component replacement problems that can be analyzed using a stochastic model are provided in Table 2.2.

Determination of replacement decisions for probabilistically failing equipment involves a problem of decision making under one main source of uncertainty: it is impossible to predict with certainty when a failure will occur, or more

TABLE 2.1
Examples of Replaceable Components that Deteriorate Deterministically

Fuel filter (automobile): as filter ages the rate of fuel consumption increases
V-belt on autowrapper used in candy plant to wrap tablets: productivity decreases as V-belt slackens
Brake and clutch module on stamping press: productivity decreases as module ages
Paper mill felt: productivity decreases as felt ages
Molds for glass production: productivity decreases as molds age

TABLE 2.2
Examples of Stochastically Failing Components

Lightbulbs
Displacement diaphragms on a food packaging line
Air conditioning (a/c) charge adapter nose seal on a/c evacuate and fill equipment in automanufacture
Top and bottom guide apron cylinders in a steel mill
Fuel injectors on the main propulsion diesel engine onboard a ship

generally, when the transition from one state of the equipment to another will occur. A further source of uncertainty is that it may be impossible to determine the state of equipment, either good, failed, or somewhere between, unless definite maintenance action is taken, such as inspection. This aspect of uncertainty is highly relevant to equipment, often termed protective devices, used in emergency situations. An example of such a protective device is a pressure safety valve in an oil and gas field — if it is dormant, waiting to come into service when an unacceptable pressure level occurs. Its condition can only be determined through an inspection. These problems will be covered in Chapter 3.

In the probabilistic problems of this chapter it will be assumed that there are only two possible conditions of the equipment, good and failed, and that the condition is always known. This is not unreasonable since, for example, with continuously operating equipment producing some form of goods, we soon know when the equipment reaches the failed state because items may be produced outside specified tolerance limits or the equipment may cease to function.

In determining when to perform a replacement, we are interested in the sequence of times at which the replacement actions should take place. Any sequence of times is a replacement policy, but what we are interested in determining are optimal replacement policies, that is, ones that maximize or minimize some criterion, such as profit, total cost, and downtime, or ensure that a specified safety or environmental criterion is not exceeded.

In many of the models of component replacement problems presented in this chapter it will be assumed, which applies in many cases, that the replacement action returns the equipment to the "as new" condition, thus continuing to provide exactly the same services as the equipment that has just been replaced when it was new. By making this assumption, we are implying that various costs, failure distributions, and so on, used in the analysis do not change from one replacement to the next. An exception to this assumption will be problems where the item being replaced is not replaced by one that can be considered statistically as good as new. If this is the case, we are often dealing with a repairable component: such problems will be addressed in Section 2.9.3.

Throughout this chapter, maintenance actions such as overhaul and repair can be considered to be equivalent to replacement, provided it is reasonable to assume that such actions also return equipment to the as-new condition. In practice, this is often a reasonable assumption, and hence the following models can often be used to analyze overhaul/repair problems. If it is not reasonable to make such an assumption, then models introduced in Section 2.9.3, along with the model associated with condition-based maintenance in Chapter 3, may help.

Section 2.2 addresses a common deterministic component replacement problem. Stochastic problems are covered in Sections 2.3 to 2.9.

2.2 OPTIMAL REPLACEMENT TIMES FOR EQUIPMENT WHOSE OPERATING COST INCREASES WITH USE

2.2.1 STATEMENT OF PROBLEM

Some equipment operates with excellent efficiency when it is new, but as it ages its performance deteriorates. An example is the air filter in an automobile. When new, there is good gasoline consumption, but as the air filter gets dirty, the gasoline consumption per kilometer increases. The question then is: When on the increasing cost trend is it economically justifiable to replace the air filter, thus reducing the operating cost of the automobile? In general, replacements cost money in terms of materials and wages, and a balance is required between the money spent on replacements and savings obtained by reducing the operating cost. Thus, we wish to determine an optimal replacement policy that will minimize the sum of operating and replacement costs per unit time.

When dealing with optimization problems, in general, we wish to optimize some measure of performance over a long period. In many situations, this is equivalent to optimizing the measure of performance per unit time. This approach is easier to deal with mathematically when compared to developing a model for optimizing a measure of performance over a finite horizon.

The cost conflicts and associated optimization problem are illustrated in Figure 2.3. It should be stressed that this class of problem can be termed short-term deterministic since the magnitude of the interval between replacements is weeks or months, rather than years. If the interval between replacements was measured in years, then the fact that money changes in value over time would need to be taken into account in the analysis. Such problems can be termed long-term replacement and are dealt with in Chapter 4.

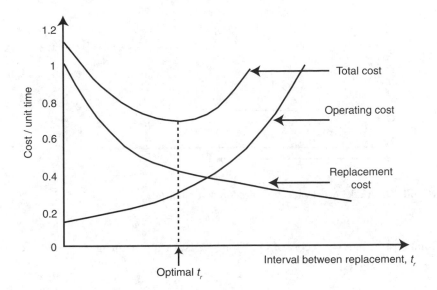

FIGURE 2.3 Short-term deterministic optimization.

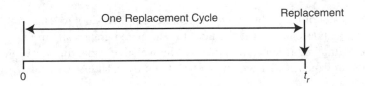

FIGURE 2.4 A replacement cycle.

2.2.2 CONSTRUCTION OF MODEL

1. $c(t)$ is the operating cost per unit time at time t after replacement.
2. C_r is the total cost of a replacement.
3. The replacement policy is to perform replacements at intervals of length t_r. The policy is illustrated in Figure 2.4.
4. The objective is to determine the optimal interval between replacements to minimize the total cost of operation and replacement per unit time.

The total cost per unit time $C(t_r)$, for replacement at time t_r, is

$$C(t_r) = \text{total cost in interval } (0, t_r)/\text{length of interval}$$

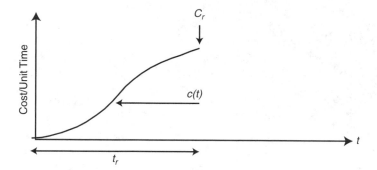

FIGURE 2.5 Model development: short-term deterministic.

Total cost in interval = cost of operating + cost of replacement = $\displaystyle\int_0^{t_r} c(t)dt + C_r$

$$C(t_r) = \frac{1}{t_r}\left[\int_0^{t_r} c(t)dt + C_r\right] \qquad (2.1)$$

This is a model of the problem relating replacement interval t_r to total cost per unit time $C(t_r)$, and development of the model is illustrated graphically in Figure 2.5.

The optimal replacement interval t_r is that value of t_r that minimizes the right-hand side of Equation 2.1 which can be shown by calculus to occur when

$$c(t_r) = C(t_r)$$

Thus, the optimal replacement time is when the current operating cost rate is equal to the average total cost per unit time. In other words, the optimal time to replace is when the marginal cost equals the average cost.

In fact, if the trend in operating costs is linear, $c(t) = a + bt$, then the optimal replacement interval t^* is

$$t^* = \sqrt{\frac{2C_r}{b}}$$

To use the equation $c(t_r) = C(t_r)$ requires that the trend in operating costs be an increasing function, which in practice is a very reasonable assumption. If that is not the case, and as time progresses the operating cost of a component becomes less, then Equation 2.1 needs to be solved using classical calculus (if the cost trend is simple); otherwise, a numerical solution will be required.

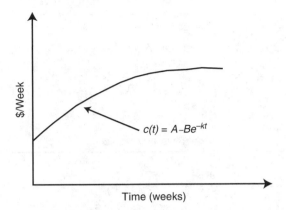

FIGURE 2.6 Exponential trend in operating costs.

If the trend in operating costs is not continuous, but discrete, then the optimal replacement time is when the next period's operating cost is equal to or greater than the current average cost of replacement to that time. In other words, replace when the marginal operating cost is greater than the average cost to date.

2.2.3 NUMERICAL EXAMPLE

1. The trend in operating cost for an item is of the form

$$c(t) = A - B \exp[-kt]$$

   where $A = \$100$, $B = \$80$, and $k = 0.21$/week.
   This trend is illustrated in Figure 2.6. Note: $A - B \geq 0$ may be interpreted as the operating cost per unit time if no deterioration occurs. k is a constant describing the rate of deterioration.
2. C_r, the total cost of a replacement, is $100.
   Thus,

$$C(t_r) = \frac{1}{t_r}\left[\int_0^{t_r}(100 - 80\exp[-0.21t])dt + 100\right]$$

In this case, an analytical solution (closed form) using the result $c(t_r) = C(t_r)$ cannot be obtained. A numerical solution is required, or discrete time can be used. Evaluation of the above model for different values of t_r is given in Table 2.3, indicating that the optimal value of t_r is at 5 weeks.

TABLE 2.3
Optimal Replacement Age

t_r	1	2	3	4	5	6	7
$C(t_r)$	127.8	84.7	74.0	70.9	70.5	71.5	72.5

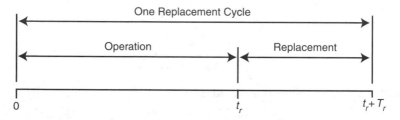

FIGURE 2.7 Replacement cycle.

2.2.4 FURTHER COMMENTS

In construction of the model in this section, the time required to effect a replacement has not been included. This replacement time can be catered for without difficulty (see Figure 2.7 and Equation 2.2, which is the appropriate model).

In practice, it is often not unreasonable to disregard the replacement time since it is usually small when compared to the interval between the replacements. Any costs, such as production losses incurred due to the duration of the replacement, need to be incorporated into the cost of the replacement action.

$$C(t_r) = \frac{\int_0^{t_r} c(t)dt + C_r}{t_r + T_r} \tag{2.2}$$

Models have now been developed whereby, for particular assumptions, the optimal interval between replacements can be obtained. In practice, there may be considerable difficulty in scheduling replacements to occur at their optimal time, or in obtaining the values of some of the parameters required for the analysis. To further assist the engineer in deciding what an appropriate replacement policy should be, it is usually useful to plot the total cost/unit time curve (Figure 2.8). The advantage of the curve is that, along with giving the optimal value of t_r, it shows the form of the total cost around the optimum. If the curve is fairly flat around the optimum, it is not really very important that the engineer should plan for the replacements to occur exactly at the optimum, thus giving some leeway in scheduling the work. Thus, in Figure 2.8 a replacement interval (t_r) with a

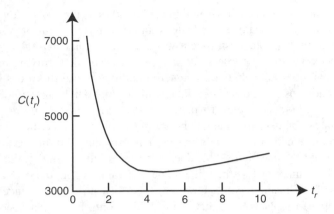

FIGURE 2.8 Form of total cost curve.

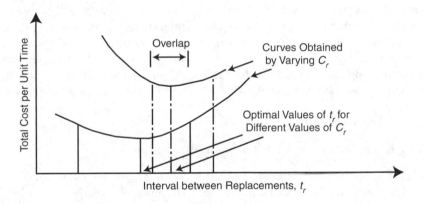

FIGURE 2.9 Sensitivity analysis of cost function.

value somewhere between 3.5 and 6 weeks does not influence greatly the total cost. Of course, if the total cost curve is not fairly flat around the optimum but rising rapidly on both sides, then the optimal interval should be adhered to if at all possible.

If there is uncertainty about the value of the particular parameter required in the analysis — say we are not sure what the replacement cost is — then evaluation of the total cost curve for various values of the uncertain parameter, and noting the effect of this variation on the optimal solution, often goes a long way toward deciding what policy should be adopted and if the particular parameter is important from a solution viewpoint. For example, varying C_r in Equation 2.1 may produce curves similar to Figure 2.9, which demonstrate, in this instance, that although C_r is varied, it does not greatly influence the optimal values of t_r. In fact, there is an overlap, which indicates a good solution independent of the true value of C_r (provided this value is within the bounds specified by the two curves). If changes in C_r drastically altered the solution from the point of replacement interval and minimal total cost, then it would be clear that a careful study would be required to identify the true value of C_r to be used when solving the model. (For example, does C_r include only material and labor costs? Or does it include lost production costs? Or costs associated with having to use a less efficient plant, overtime, or contractors, etc., to make up losses incurred resulting from the replacement?) The decision that can be taken (in this case regarding the interval between replacements) essentially may remain constant within the uncertainty region checked by sensitivity. This does not necessarily mean that the true total costs will have more or less the same numerical value within the overlap region. From a decision-making point, however, this does not matter, since it is the interval between replacements that is under the control of the decision maker. The total costs are a consequence of the decision taken.

Thus, it is seen that sensitivity checking gives guidance on what information is important from a decision-making viewpoint and, consequently, what information should be gathered in a data collection scheme. The statement "garbage in = garbage out" (GIGO), which is frequently made with reference to data requirements of quantitative techniques, is also demonstrated to be not necessarily correct. The validity of GIGO does depend on the sensitivity of the solution to particular garbage. Note, therefore, that GI does not necessarily equal GO, and so our information requirements for the use of quantitative techniques may not be as severe as is often claimed.

2.2.5 APPLICATIONS

2.2.5.1 Replacing the Air Filter in an Automobile

What is the economic replacement time for the air filter in an automobile?

The purchase price of an air filter is $80. The automobile driver travels 2,000 km/month.

Gasoline costs $0.75/l. When the air filter is new, then during the first month of operation the automobile's performance is 15 km/l; thus, the first month's operating cost is $100.00. As the filter ages, there is deterioration in the amount of kilometers that can be driven using 1 liter of gasoline. The deterioration trend is given in Table 2.4.

Using Equation 2.1, in discrete form, we obtain Table 2.5, from which it is seen that the optimal replacement age is 4 months, and the associated cost per month is $131.88. The associated graph of cost per month vs. time is provided

TABLE 2.4
Deteriorating Trend in Distance Traveled

Age (months)	1	2	3	4	5
Km/l	15	14	13	12	10

TABLE 2.5
Optimal Filter Change-Out Time

T (month)	c(t) ($)	C(t) ($/month)
1	100.00	(100 + 80)/1 = 180
2	107.13	(100 + 107.13 + 80)/2 = 143.57
3	115.38	(100 + 107.13 + 115.38 + 80)/3 = 134.17
4	125.00	(100 + 107.13 + 115.38 + 125 + 80)/4 = 131.88
5	150.00	(100 + 107.13 + 115.38 + 125 + 150 + 80)/5 = 135.50

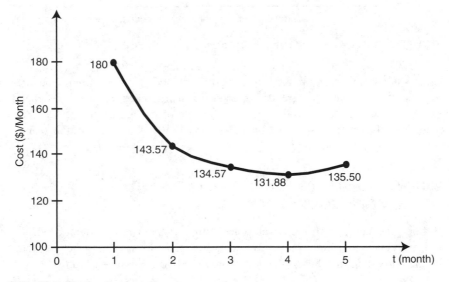

FIGURE 2.10 Graph of total cost per month.

in Figure 2.10, which includes the calculation showing the use of the optimizing criterion $c(t) = C(t_r)$ when the trend in operating cost is discretized.

Therefore, replace at the end of month 4, since next period's operations and maintenance (O&M) cost, $c(t = 5)$, is greater than the average cost to date ($131.88).

2.2.5.2 Overhauling Boiler Plant

The replacement problem we have been discussing is similar to a problem associated with a boiler plant. Through use, the heat transfer surfaces within the boiler become less efficient, and to increase their efficiency, they can be cleaned. Cleaning thus increases the rate of heat transfer, and less fuel is required to produce a given amount of steam. However, due to deterioration of other parts of the boiler plant, the trend in operating cost is not constant after each cleaning operation (equivalent to a replacement), but follows a trend similar to that of Figure 2.11.

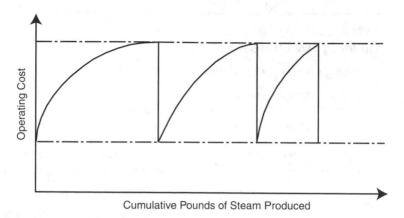

FIGURE 2.11 Operating cost trend: boiler plant.

Thus, k illustrated in Figure 2.6 is no longer constant, but varies from replacement to replacement. That is, the trend in operating cost after each replacement depends on the amount of steam produced up to the date of the replacement. A detailed study of this problem is given by Davidson (1970), who analyzed it using a dynamic programming model.

2.3 STOCHASTIC PREVENTIVE REPLACEMENT: SOME INTRODUCTORY COMMENTS

Before proceeding with the development of component replacement models, it is important to note that preventive replacement actions, that is, actions taken before equipment reaches a failed state, require two necessary conditions:

1. The total cost of the replacement must be greater after failure than before (if cost is the appropriate criterion; otherwise, an appropriate criterion, such as downtime, is substituted in place of cost). This may be caused by a greater loss of production since replacement after failure is unplanned or failure of one piece of plant may cause damage to other equipment. For example, replacement of a piston ring in an automobile engine before failure of the ring may only involve the cost of a piston ring plus a labor charge, whereas after failure its replacement cost may also include the cost of a cylinder rebore.
2. The hazard rate of the equipment must be increasing. To illustrate this point, we may consider equipment with a constant hazard rate. That is, failures occur according to the negative exponential distribution or, equivalently, the Weibull distribution, where the shape parameter $\beta = 1.0$. When this is the case, replacement before failure does not affect the probability that the equipment will fail in the next instant, given that it is good now. Consequently, money and time are wasted if

preventive replacement is applied to equipment that fails according to the negative exponential distribution. Obviously, when equipment fails according to the hyperexponential distribution or the Weibull distribution whose β value is less than 1.0, its hazard rate is decreasing and again component preventive replacement should not be applied. Examples of components where a decreasing hazard rate has been identified include quartz crystals, medium- and high-quality resistors, and capacitors and solid-state devices such as semiconductors and integrated circuits (TEAM, 1976).

In practice, it is useful to appreciate that the hazard rate of equipment must be increasing before preventive replacement is worthwhile. Very often, when equipment frequently breaks down, the immediate reaction of the maintenance professional is that the level of preventive replacement should be increased. If the failure distribution of the components being replaced had been identified through conducting a Weibull analysis (see Appendix 2), it would be clear whether such preventive replacement was applicable. It may well be that the appropriate procedure is to allow the item to fail before performing a replacement, and this decision can be made simply by obtaining statistics relevant to the equipment and does not involve the construction and solution of a model to analyze the problem. If improved system reliability is required, then a redesign is required. This may include introducing redundant components.

Note, however, that preventive maintenance of a general nature, which does not return equipment to the as-new condition, may be appropriate for equipment that is subject to a constant hazard rate. Determination of the best level of such preventive work will be covered in Chapter 3 in a problem relating to determination of the optimal frequency of inspection and associated minor maintenance of complex equipment.

2.4 OPTIMAL PREVENTIVE REPLACEMENT INTERVAL OF ITEMS SUBJECT TO BREAKDOWN (ALSO KNOWN AS THE GROUP OR BLOCK POLICY)

2.4.1 STATEMENT OF PROBLEM

An item, sometimes termed a line replaceable unit (LRU) or part, is subject to sudden failure, and when failure occurs, the item has to be replaced. Since failure is unexpected, it is not unreasonable to assume that a failure replacement is more costly than a preventive replacement. For example, a preventive replacement is planned and arrangements are made to perform it without unnecessary delays, or perhaps a failure may cause damage to other equipment. In order to reduce the number of failures, preventive replacements can be scheduled to occur at specified intervals. However, a balance is required between the amount spent on the preventive replacements and their resulting benefits, that is, reduced failure replacements. The conflicting cost consequences and their resolution by identifying the total cost curve are illustrated in Figure 2.12.

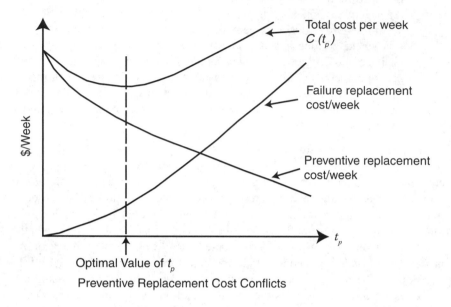

Preventive Replacement Cost Conflicts

FIGURE 2.12 Optimal replacement time.

In this section it will be assumed, not unreasonably, that we are dealing with a long period over which the equipment is to be operated and the intervals between the preventive replacements are relatively short. When this is the case, we need consider only one cycle of operation and develop a model for one cycle. If the interval between the preventive replacements is long, it would be necessary to use a discounting approach, and the series of cycles would have to be included in the model (see Chapter 4) to take into account the time value of money.

The replacement policy is one where preventive replacements occur at fixed intervals of time; failure replacements occur whenever necessary. We want to determine the optimal interval between the preventive replacements to minimize the total expected cost of replacing the equipment per unit time.

2.4.2 CONSTRUCTION OF MODEL

1. C_p is the total cost of a preventive replacement.
2. C_f is the total cost of a failure replacement.
3. $f(t)$ is the probability density function of the item's failure times.
4. The replacement policy is to perform preventive replacements at constant intervals of length t_p, irrespective of the age of the item, and failure replacements occur as many times as required in interval $(0, t_p)$. The policy is illustrated in Figure 2.13.

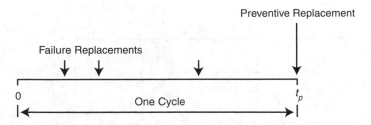

FIGURE 2.13 Replacement cycle: constant–interval policy.

5. The objective is to determine the optimal interval between preventive replacements to minimize the total expected replacement cost per unit time.

The total expected cost per unit time for preventive replacement at intervals of length t_p denoted $C(t_p)$ is

$$C(t_p) = \text{total expected cost in interval } (0, t_p)/\text{length of interval}$$

Total expected cost in interval $(0, t_p)$

$$= \text{cost of a preventive replacement} + \text{expected cost of failure replacements}$$
$$= C_p + C_f H(t_p)$$

where $H(t_p)$ is the expected number of failures in interval $(0, t_p)$.

Length of interval $= t_p$

Therefore,

$$C(t_p) = \frac{C_p + C_f H(t_p)}{t_p} \tag{2.3}$$

This is a model of the problem relating replacement interval t_p to total cost $C(t_p)$.
    Differentiating the right-hand side of Equation 2.3 with respect to t_p and equating it to zero gives the optimized result:

$$t_p\, h(t_p) - H(t_p) = C_p/C_f$$

where $h(t_p)$ is the derivative of $H(t_p)$ and termed the renewal density: $\int_0^{t_p} h(t)\, dt = H(t_p)$. See Section 2.4.3 for the derivation of $H(t)$.
    A numerical solution to Equation 2.3 will be illustrated by an example in Section 2.4.4.

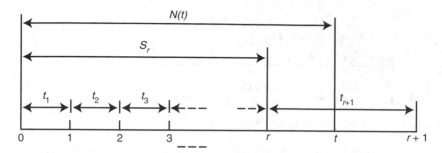

FIGURE 2.14 Establishing $H(t)$.

Before proceeding with the example, we will illustrate a procedure to determine $H(t_p)$, the expected number of failures in an interval of length t_p.

2.4.3 Determination of $H(t)$

2.4.3.1 Renewal Theory Approach

With reference to Figure 2.14, we may define the following terms:

$N(t)$ is the number of failures in interval $(0, t)$.
$H(t)$ is the expected number of failures in interval $(0, t) = E[N(t)]$, where
$E[\cdot]$ denotes expectation.
t_1, t_2, etc., are the intervals between failures.
S_r is the time up to the rth failure $= t_1 + t_2 + \ldots + t_r$.

Now the probability of $N(t) = r$ is the probability that t lies between the rth and $(r + 1)^{th}$ failure. This is obtained as follows:

$$P[N(t) < r] = 1 - F_r(t)$$

where $F_r(t)$ is the cumulative distribution function of S_r:

$$P[N(t) > r] = F_{r+1}(t)$$

Now,

$$P[N(t) < r] + P[N(t) = r] + P[N(t) > r] = 1$$

Therefore,

$$P[N(t) = r] = F_r(t) - F_{r+1}(t)$$

The expected value of $N(t)$ is then

$$H(t) = \sum_{r=0}^{\infty} rP[N(t) = r] = \sum_{r=0}^{\infty} r[F_r(t) - F_{r+1}(t)]$$

$$H(t) = \sum_{r=1}^{\infty} F_r(t) \tag{2.4}$$

On taking Laplace transforms[†] of both sides of Equation 2.4 we get

$$H^*(s) = \frac{f^*(s)}{s[1 - f^*(s)]} \tag{2.5}$$

The problem is then to determine $H(t)$ from $H^*(s)$. This is done by determining $f(t)$ from $f^*(s)$, a process termed inversion. Inversion is usually done by reference to tables giving Laplace transforms of functions and giving $f(t)$ corresponding to common forms of $f^*(s)$.

EXAMPLE

If $f(t) = \lambda e^{-\lambda t}$, then from the tables $f^*(s) = \lambda / (\lambda + s)$. From Equation 2.5,

$$H^*(s) = \frac{\lambda / (\lambda + s)}{s[1 - \lambda / (\lambda + s)]} = \lambda / s^2$$

From the tables, the function corresponding to $1/s^2$ is t, and so

$$H(t) = \lambda t$$

In practice, $H^*(s)$ can only be inverted in simple cases. However, if t is large (tending to infinity),

$$H(t) \approx \frac{t}{\mu} + \frac{\sigma^2 - \mu^2}{2\mu^2} \tag{2.6}$$

where μ and σ^2 are the mean and variance of $f(t)$.

[†]If $f(t)$ is the probability density function of a nonnegative random variable T, the Laplace transform $f^*(s)$ is defined by

$$f^*(s) = E\exp[-sT] = \int_0^{\infty} \exp[-st]f(t)dt.$$

The main importance of Laplace transforms in renewal theory is in connection with the sums of independent random variables. For further details of renewal theory, see Cox (1962).

EXAMPLE

An item fails according to the normal distribution with $\mu = 5, \sigma^2 = 1$. If the interval between preventive replacements is $t = 1,000$ weeks, then from Equation 2.6,

$$H(1000) \approx \frac{1000}{5} + \frac{1-25}{50} = 199 \cdot 5 \text{ failures}$$

Of course, we do not expect to get large numbers of failures between preventive replacements (if we do, we are not doing preventive replacement), and so Equation 2.6 is not appropriate and therefore we would need to use Equation 2.5. In order to avoid possible difficulties of inverting $H^*(s)$, a discrete approach is usually adopted to determine $H(t)$.

2.4.3.2 Discrete Approach

Figure 2.15 illustrates the case where there are 4 weeks between preventive replacements. Then $H(4)$ is the expected number of failures in interval $(0, 4)$, starting with new equipment.

When we start at time zero, the first failure (if there is one) will occur during either the first, second, third, or fourth week of operation. Keeping this fact in mind, we get the following:

$H(4)$ = number of expected failures that occur in interval $(0, 4)$ when the first failure occurs in the first week × probability of the first failure occurring in interval $(0, 1)$

+ Number of expected failures that occur in interval $(0, 4)$ when the first failure occurs in the second week × probability of the first failure occurring in interval $(1, 2)$

+ Number of expected failures that occur in interval $(0, 4)$ when the first failure occurs in the third week × probability of the first failure occurring in interval $(2, 3)$

+ Number of expected failures that occur in interval $(0, 4)$ when the first failure occurs in the fourth week × probability of the first failure occurring in interval $(3, 4)$.

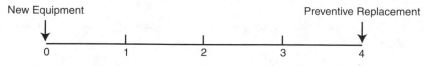

New Equipment   Preventive Replacement

0   1   2   3   4

FIGURE 2.15 Establishing $H(t)$: discrete approach.

Assume that not more than one failure can occur in any weekly interval. This is not restrictive since the length of each interval can be made as short as desired. If this is the case, then

Number of expected failures that occur in interval (0, 4) when the first failure occurs in the first week

= the failure that occurred in the first week + the expected number of failures in the remaining 3 weeks

= 1 + H(3)

Note we use $H(3)$ since we have new equipment as a result of replacing the failed component in the first week, and we then have 3 weeks to go before the preventive replacement occurs. By definition, the expected number of failures in the remaining 3 weeks, starting with the new equipment, is $H(3)$.

The probability of the first failure occurring in the first week $= \int_{0}^{1} f(t)dt$ in the first week. Similarly, in consequence of the first failure occurring in the second, third, or fourth weeks,

$$H(4) = [1 + H(3)]\int_{0}^{1} f(t)dt + [1 + H(2)]\int_{1}^{2} f(t)dt + [1 + H(1)]\int_{2}^{3} f(t)dt + [1 + H(0)]\int_{3}^{4} f(t)dt$$

Obviously, $H(0) = 0$. That is, with zero weeks to go, the expected number of failures is zero.

Tidying up the above equation we get

$$H(4) = \sum_{i=0}^{3} [1 + H(3-i)]\int_{i}^{i+1} f(t)dt$$

with $H(0) = 0$.

In general,

$$H(T) = \sum_{i=0}^{T-1} [1 + H(T-i-1)]\int_{i}^{i+1} f(t)dt, T \geq 1 \qquad (2.7)$$

with $H(0) = 0$.

Equation 2.7 is termed a recurrence relation. Since we know that $H(0) = 0$, we can get $H(1)$, then $H(2)$, then $H(3)$, and so on, from Equation 2.7.

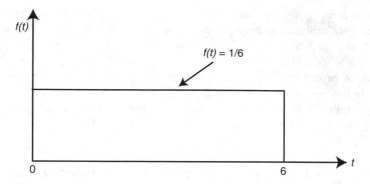

FIGURE 2.16 Uniform distribution.

EXAMPLE

Assume $f(t) = \dfrac{1}{6}, 0 \le t \le 6$. This is illustrated in Figure 2.16, which is termed a uniform or rectangular distribution. Determine the expected number of failures if preventive replacements occur every 2 weeks.

In this case, we want $H(2)$. From Equation 2.7,

$$H(2) = \sum_{i=0}^{1} [1 + H(1-i)] \int_{i}^{i+1} f(t)\,dt$$

$$= [1 + H(1)] \int_{0}^{1} f(t)\,dt + [1 + H(0)] \int_{1}^{2} f(t)\,dt$$

Now,

$$H(0) = 0$$

$$H(1) = [1 + H(0)] \int_{0}^{1} f(t)\,dt \text{ from Equation 2.7}$$

$$= \int_{0}^{1} \frac{1}{6}\,dt = \frac{1}{6}$$

$$H(2) = \left(1 + \frac{1}{6}\right) \int_{0}^{1} \frac{1}{6}\,dt + (1+0) \int_{1}^{2} \frac{1}{6}\,dt$$

$$= \frac{7}{6} \times \frac{1}{6} + 1 \times \frac{1}{6} = \frac{13}{36}$$

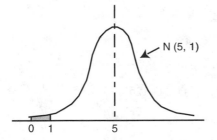 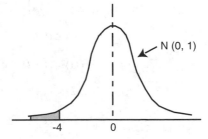

FIGURE 2.17 Use of the normal distribution.

TABLE 2.6
Optimal Preventive Replacement Interval

t_p	1	2	3	4	5	6
$C(t_p)$	5.00	2.51	1.74	1.65	2.00	2.24

2.4.4 NUMERICAL EXAMPLE

Given $C_p = \$5$, $C_f = \$10$, determine the optimal replacement interval for equipment subject to the replacement strategy of Section 2.4. Failures occur according to the normal distribution with mean = 5 weeks, standard deviation = 1 week. (See Figure 2.17.) From Equation 2.3 we have

$$C(t_p) = \frac{5 + 10H(t_p)}{t_p}$$

Values of $C(t_p)$ for various values of t_p are given in Table 2.6, from which it is seen that the optimal replacement policy is to perform preventive replacements every 4 weeks.

Sample calculation for $t_p = 2$ weeks:

$$H(2) = [1 + H(1)] [\Phi(-4) - \Phi(-5)] + [1 + H(0)] [\Phi(-3) - \Phi(-4)]$$

(See Appendix 1 for guidance on reading the table of the standardized normal distribution.)

From the tables, $\Phi(-4) \approx 0$ and $\Phi(-5) \approx 0$, and now

$$\Phi(-3) - \Phi(-4) = 0.00135 - 0 = 0.00135$$

$$H(0) = 0$$

$$H(1) = [1 + H(0)] \times 0 = 0$$

$$H(2) = (1 + 0) \times 0 + (1 + 0) \times 0.00135 = 0.00135$$

Therefore,

$$C(2) = (5 + 10 \times 0.00135) / 2 = \$2.51 \text{ per week}$$

2.4.5 FURTHER COMMENTS

In this example, no account was taken of the time required to perform failure and preventive replacements since they were considered to be very short (hours or days), compared to the mean time between replacement of an item, which may be measured in weeks or months. When necessary, the replacement durations can be incorporated into the replacement model, as is required when the goal is the minimization of total downtime or, equivalently, the maximization of item availability. This will be presented in Section 2.6. Of course, any costs that are incurred due to the replacement stoppages need to be included as part of C_p, the total cost of a preventive replacement, and C_f, the total cost of a failure replacement.

In this section, when establishing the optimal preventive replacement interval, we used the term *time*. In practice, what we measure is one indicator that is used to monitor the health of an item. Calendar time is perfectly acceptable if an item's utilization is constant, but if that is not the case, then a better variable to measure the working age of the item needs to be used, such as operating hours, weight of material processed, cycles of operation, and so on. The key issue with component preventive replacement that is being addressed in this chapter is that only one variable is being used to estimate the health of the item as described by its probability of failure. Later in Chapter 3, when we deal with condition-based maintenance, we will estimate the health of an item through taking into account not only age, but also measurements that are acquired at the time of condition monitoring.

In Chapter 1, the methodology of RCM was discussed. It should be noted that one of the maintenance tactics identified that results from employing RCM is termed the time-based discard decision. This section has presented a model that can be used to establish the optimal time-based discard decision if the goal is a constant-interval preventive replacement policy.

2.4.6 AN APPLICATION: OPTIMAL REPLACEMENT INTERVAL
FOR A LEFT-HAND STEERING CLUTCH

In an open-pit mining operation the current policy was to replace the left-hand steering clutch on a piece of mobile equipment only when it failed. In this application there was a fleet of six identical machines, all operating in the same environment. The fleet had experienced seven failures. When the study was being undertaken, all six machines were operating in the mine site. To increase the sample size, data on the present ages of the clutches on the six currently operating machines were obtained, and thus the data available for analysis were seven failures and six suspensions. Using the procedure described in Appendix 2 for blending together failure and suspension data, the Weibull shape parameter β was estimated as 1.79, and the mean time to failure was estimated as 6500 hours. This indicates that there is an increasing probability of the clutch failing as it ages since β is greater than 1.0.

Determining the optimal preventive replacement age to minimize total cost requires that the costs be obtained. In this case, the total cost of a preventive replacement was obtained by adding the cost of labor (16 hours — 2 people each at 8 hours), parts, and equipment out-of-service cost (8 hours). The cost of a failure replacement was obtained from adding the labor cost (24 hours — 2 people at 12 hours), parts, and equipment out-of-service cost (12 hours).

While the cost consequence associated with a failure replacement was greater than that for a preventive replacement, it was not sufficiently large to warrant changing the current policy of replace only on failure (R-o-o-F). But at least the mining operation had an evidence-based decision. As the maintenance superintendent subsequently said, "A run-to-failure policy was a surprising conclusion since the clutch was exhibiting wear-out characteristics. However, the economic considerations did not justify preventive replacement according to a fixed-time maintenance policy."

2.5 OPTIMAL PREVENTIVE REPLACEMENT AGE OF AN ITEM SUBJECT TO BREAKDOWN

2.5.1 STATEMENT OF PROBLEM

This problem is similar to that of Section 2.4, except that instead of making preventive replacements at fixed intervals, with the possibility of performing a preventive replacement shortly after a failure replacement, the time at which the preventive replacement occurs depends on the age of the item. When failures occur, failure replacements are made. When this occurs, the time clock is reset to zero, and the preventive replacement occurs only when the item has been in use for the specified period.

Again, the problem is to balance the cost of the preventive replacements against their benefits, and we do this by determining the optimal preventive replacement age for the item to minimize the total expected cost of replacements per unit time.

2.5.2 CONSTRUCTION OF MODEL

1. C_p is the total cost of a preventive replacement.
2. C_f is the total cost of a failure replacement.

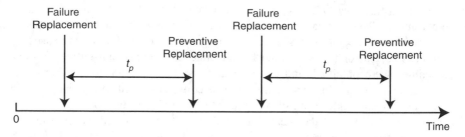

FIGURE 2.18 Replacement cycles: age-based policy.

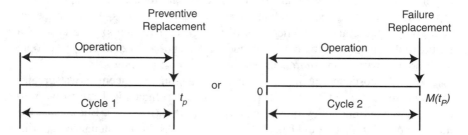

FIGURE 2.19 Possible replacement cycles: age-based replacement.

3. $f(t)$ is the probability density function of the failure times of the item.
4. The replacement policy is to perform a preventive replacement when the item has reached a specified age, t_p, plus failure replacements when necessary. This policy is illustrated in Figure 2.18.
5. The objective is to determine the optimal replacement age of the item to minimize the total expected replacement cost per unit time.

In this problem, there are two possible cycles of operation: one cycle being determined by the item reaching its planned replacement age, t_p, and the other being determined by the equipment ceasing to operate due to a failure occurring before the planned replacement time. These two possible cycles are illustrated in Figure 2.19.

The total expected replacement cost per unit time, $C(t_p)$, is

$$C(t_p) = \frac{\text{Total expected replacement cost per cycle}}{\text{Expected cycle length}}$$

Note: We are obtaining the expected cost per unit time as a ratio of two expectations. This is acceptable in many replacement problems since it has been shown (Smith, 1955) that

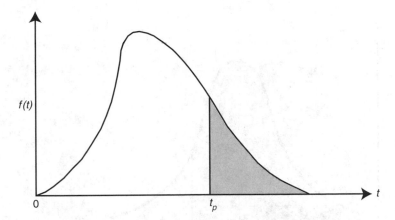

FIGURE 2.20 Item failure distribution.

$$\lim_{t\to\infty} \frac{K(t)}{t} = \frac{\text{Expected cost per cycle}}{\text{Expected cycle length}}$$

where $K(t)$ is the cumulative expected cost due to a series of cycles in an interval $(0, t)$. $K(t)/t$ is the expected cost per unit time.

Total expected replacement cost per cycle
    = cost of a preventive cycle × probability of a preventive cycle
       + cost of a failure cycle × probability of a failure cycle
    = $C_p R(t_p) + C_f \times [1 - R(t_p)]$

Remember: If $f(t)$ is as illustrated in Figure 2.20, then the probability of a preventive cycle equals the probability of failure occurring after time t_p, that is, equivalent to the shaded area, which is denoted as $R(t_p)$. (Refer to Appendix 1 for a discussion of the reliability function.)

The probability of a failure cycle is the probability of a failure occurring before time t, which is the unshaded area of Figure 2.20. Since the area under the curve equals unity, then the unshaded area is $[1 - R(t)]$.

Expected cycle length
    = length of a preventive cycle × probability of a preventive cycle
       + expected length of a failure cycle × probability of a failure cycle
    = $t_p \times R(t_p)$ + (expected length of a failure cycle) × $[1 - R(t_p)]$

To determine the expected length of a failure cycle, consider Figure 2.21.

The mean time to failure of the complete distribution is $\int_{-\infty}^{\infty} t f(t) dt$, which for the

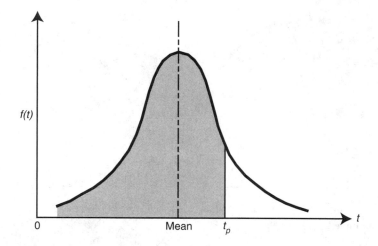

FIGURE 2.21 Estimating the mean of a truncated distribution.

normal distribution equals the mode (peak) of the distribution. If a preventive replacement occurs at time t_p, then the mean time to failure is the mean of the shaded portion of Figure 2.21, since the unshaded area is an impossible region for failures. The mean of the shaded area is $\displaystyle \int_{-\infty}^{t_p} \frac{tf(t)dt}{1-R(t_p)}$, denoted as $M(t_p)$.

Therefore,

$$\text{Expected cycle length} = t_p \times R(t_p) + M(t_p) \times [1 - R(t_p)]$$

$$C(t_p) = \frac{C_p \times R(t_p) + C_f \times [1 - R(t_p)]}{t_p \times R(t_p) + M(t_p) \times [1 - R(t_p)]} \qquad (2.8)$$

This is now a model of the problem relating replacement age t_p to total expected replacement cost per unit time.

Note: There is no simple solution to Equation 2.8, as there is for the constant-interval model, Equation 2.3. However, Equation 2.8 can be simplified to

$$C(t_p) = \frac{C_p \times R(t_p) + C_f \times [1 - R(t_p)]}{t_p \times R(t_p) + \displaystyle\int_{-\infty}^{t_p} tf(t)dt}$$

TABLE 2.7
Optimal Preventive Replacement Age

t_p	1	2	3	4	5	6
$C(t_p)$	5.00	2.50	1.70	1.50	1.63	1.87

2.5.3 NUMERICAL EXAMPLE

Using the data of the example in Section 2.4.4, determine the optimal replacement age of the equipment. Equation 2.8 becomes

$$C(t_p) = \frac{5 \times R(t_p) + 10 \times [1 - R(t_p)]}{t_p \times R(t_p) + \int\limits_{-\infty}^{t_p} t f(t) dt}$$

For various values of t_p, the corresponding values of $C(t_p)$ are given in Table 2.7, from which it is seen that the optimal replacement age is 4 weeks.

Sample calculation for $t_p = 3$ weeks:

Equation 2.8 becomes

$$C(3) = \frac{5 \times R(3) + 10 \times [1 - R(3)]}{3 \times R(3) + \int\limits_{-\infty}^{t_p} t f(t) dt}$$

$$R(3) = 1 - \Phi(-2) \text{ (see Figure 2.22)}$$

$$= 0.9772$$

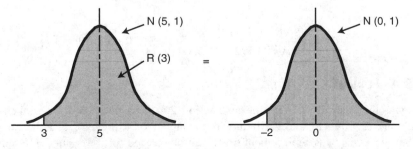

FIGURE 2.22 Calculating the probabilities for age replacement policy.

Therefore,

$$[1 - R(3)] = 1 - 0.9772 = 0.0228$$

$$\int_{-\infty}^{t_p} tf(t)dt = \frac{1}{\sigma\sqrt{2\pi}} \int_{-\infty}^{t_p} t\exp\left[\frac{-(t-\mu)^2}{2\sigma^2}\right]dt$$

and through integrating by parts we get

$$\int_{-\infty}^{t_p} tf(t)dt = -\sigma\phi\left(\frac{t_p-\mu}{\sigma}\right) + \mu\Phi\left(\frac{t_p-\mu}{\sigma}\right)$$

where $\phi(t)$ and $\Phi(t)$ are the ordinate and cumulative distribution functions at t of the standardized normal distribution, whose mean is 0 and standard deviation 1.

When $\sigma = 1, \mu = 5$, then

$$\int_{-\infty}^{3} tf(t)dt = -\phi\left(\frac{3-5}{1}\right) + 5\Phi\left(\frac{3-5}{1}\right) = -0.0540 + 5 \times 0.0228 = 0.0600$$

where 0.0540 and 0.0228 are obtained from Appendix 5 and Appendix 6, respectively.

Therefore,

$$C(3) = \frac{5 \times 0.9772 + 10 \times 0.0228}{3 \times 0.9772 + 0.0600} = \$1.70 \text{ per week}$$

2.5.4 FURTHER COMMENTS

As was the case for the example in Section 2.4, no account has been taken of the time required to make a failure replacement or a preventive replacement. When necessary, the replacement times can be catered for in the model. Section 2.6 presents a model that will include the times required to make either a failure or a preventive replacement.

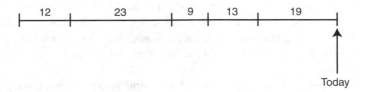

FIGURE 2.23 Historical bearing failure times.

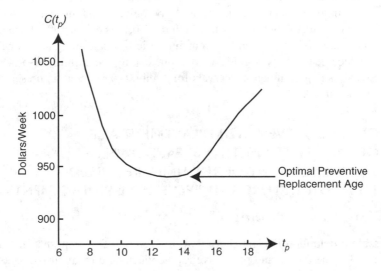

FIGURE 2.24 Establishing the optimal preventive replacement age of a bearing.

2.5.5 AN APPLICATION: OPTIMAL BEARING REPLACEMENT AGE

A critical bearing in a shaker machine in a foundry was replaced only on failure (Jardine, 1979). It was known that the cost consequence of a failure was about twice the cost of replacing the bearing under preventive conditions. Data on the most recent six failure ages were known (Figure 2.23), and from that small sample size, a Weibull analysis was undertaken to estimate the failure distribution. Best estimates using the criterion of maximum likelihood for the shape parameter (β) and the characteristic life (η) were 2.97 and 17.55 weeks, respectively. Using the age replacement model (Equation 2.8), the optimal preventive replacement age was identified as 14 weeks. Figure 2.24 is a graph of the total cost as a function of different replacement ages.

Two points are worth mentioning about this application:

1. Presenting the optimal solution to management graphically, as in Figure 2.24, is of value so management can see clearly the effect of deviating from the mathematical optimal preventive replacement age. From Figure 2.24 it is clear that a very acceptable solution is to plan to replace

the bearing somewhere between age 10 and 14 weeks. Postponing the preventive replacement age past 14 weeks is seen to drive up quickly the cost function due to the combination of risk and economics. Conversely, preventively replacing earlier than 10 weeks is seen as over-maintenance.

2. This application was one where the sample size was small, only five failures. While it is possible to obtain best estimates of the Weibull parameters — such as in this case $\beta = 2.97$ — it is also possible to place a confidence interval on the parameters. In the case of component preventive replacement, one wants to be quite confident that the confidence interval for β does not include 1.0, since if it did, it would mean that failures could be occurring strictly randomly, and the best replacement policy would be to replace only on failure (R-o-o-F). Establishing confidence intervals for Weibull parameters is presented in Section A2.4.

2.6 OPTIMAL PREVENTIVE REPLACEMENT AGE OF AN ITEM SUBJECT TO BREAKDOWN, TAKING ACCOUNT OF THE TIMES REQUIRED TO EFFECT FAILURE AND PREVENTIVE REPLACEMENTS

2.6.1 STATEMENT OF PROBLEM

The problem definition is identical to that of Section 2.5, except that instead of assuming that the failure and preventive replacements are made instantaneously, account is taken of the time required to make these replacements.

The optimal preventive replacement age of the item is again taken as that age which minimizes the total expected cost of replacements per unit time.

2.6.2 CONSTRUCTION OF MODEL

1. C_p is the total cost of a preventive replacement.
2. C_f is the total cost of a failure replacement.
3. T_p is the mean time required to make a preventive replacement.
4. T_f is the mean time required to make a failure replacement.
5. $f(t)$ is the probability density function of the failure times of the item.
6. $M(t_p)$ is the mean time to failure when preventive replacement occurs at age t_p.
7. The replacement policy is to perform a preventive replacement once the item has reached a specified age, t_p, plus failure replacements when necessary. This policy is illustrated in Figure 2.25.
8. The objective is to determine the optimal preventive replacement age of the item to minimize the total expected replacement cost per unit time.

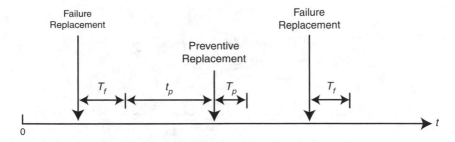

FIGURE 2.25 Age-based policy, including duration of a replacement.

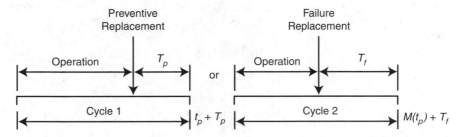

FIGURE 2.26 Age-based replacement cycles, including replacement durations.

As was the case for the problem of Section 2.4, there are two possible cycles of operation, and they are illustrated in Figure 2.26.

The total expected replacement cost per unit time, denoted $C(t_p)$, is

$$C(t_p) = \frac{\text{Total expected replacement cost per cycle}}{\text{Expected cycle length}}$$

Total expected replacement cost per cycle (as per Section 2.5) =

$$C_p \times R(t_p) + C_f[1 - R(t_p)]$$

Expected cycle length
    = length of a preventive cycle × probability of a preventive cycle
        + expected length of a failure cycle × probability of a failure cycle
= $(t_p + T_p)R(t_p) + [M(t_p) + T_f][1 - R(t_p)]$

$$C(t_p) = \frac{C_p R(t_p) + C_f[1 - R(t_p)]}{(t_p + T_p)R(t_p) + [M(t_p) + T_f][1 - R(t_p)]} \qquad (2.9)$$

This is a model of the problem relating preventive replacement age, t_p, to the total expected replacement cost per unit time.

TABLE 2.8
Optimal Preventive Replacement Age Including Replacement Times

t_p	1	2	3	4	5	6
$C(t_p)$	3.34	2.00	1.46	1.34	1.47	1.70

2.6.3 NUMERICAL EXAMPLE

For the data of Section 2.5, namely, $C_p = \$5$, $C_f = \$10$, $f(t) = N(5, 1)$, replacement times T_p and $T_f = 0.5$ week determine the optimal replacement age of the equipment.

Since $M(t_p) = \dfrac{\displaystyle\int_{-\infty}^{t_p} tf(t)dt}{1 - R(t_p)}$, Equation 2.9 becomes

$$C(t_p) = \frac{5 \times R(t_p) + 10 \times [1 - R(t_p)]}{(t_p + 0.5)R(t_p) + \displaystyle\int_{-\infty}^{t_p} tf(t)dt + 0.5 \times \left[1 - R(t_p)\right]}$$

For various values of t_p, the corresponding values of $C(t_p)$ are given in Table 2.8, from which it is seen that the optimal preventive replacement age is 4 weeks.

Sample calculation for $t_p = 3$ weeks:

Equation 2.9 becomes:

$$C(3) = \frac{5 \times R(3) + 10 \times [1 - R(3)]}{3.5 \times R(3) + 0.0600 + 0.5 \times [1 - R(3)]}$$

$$= \frac{5 \times 0.9772 + 10 \times 0.0228}{3.5 \times 0.9772 + 0.06 + 0.5 \times 0.0228}$$

$$= 5.1140 / 3.4916$$

$$= \$1.46 \text{ per week}$$

2.7 OPTIMAL PREVENTIVE REPLACEMENT INTERVAL OR AGE OF AN ITEM SUBJECT TO BREAKDOWN: MINIMIZATION OF DOWNTIME

2.7.1 STATEMENT OF PROBLEM

The problems of Sections 2.4 to 2.6 had as their objective to minimize total cost per unit time. In some cases, say due to difficulties in costing or the desire to get maximum throughput or utilization of equipment, the replacement policy required may be one that minimizes total downtime per unit time or, equivalently, maximizes availability. The problem of this section is to determine the best times at which replacements should occur to minimize total downtime per unit time. The basic conflicts are that as the preventive replacement frequency increases, there is an increase in downtime due to these replacements, but a consequence of this is a reduction of downtime due to failure replacements, and we wish to get the best balance between them.

2.7.2 CONSTRUCTION OF MODELS

1. T_f is the mean downtime required to make a failure replacement.
2. T_p is the mean downtime required to make a preventive replacement.
3. $f(t)$ is the probability density function of the failure times of the item.

2.7.2.1 Model 1: Determination of the Optimal Preventive Replacement Interval

4. The objective is to determine the optimal replacement interval t_p between preventive replacements in order to minimize total downtime per unit time. The policy is illustrated in Figure 2.27.

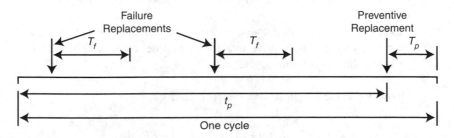

FIGURE 2.27 Downtime minimization: optimal interval.

The total downtime per unit time, for preventive replacement at time t_p, denoted as $D(t_p)$, is

$$D(t_p) = \frac{\text{Expected downtime due to failures} + \text{Downtime due to preventive replacement}}{\text{Cycle length}}$$

Downtime due to failures = number of failures in interval $(0, t_p) \times$ Time required to make a failure replacement $= H(t_p) \times T_f$

Downtime due to preventive replacement $= T_p$

Therefore,

$$D(t_p) = \frac{H(t_p)T_f + T_p}{t_p + T_p} \qquad (2.10)$$

This is a model of the problem relating replacement interval t_p to total downtime $D(t_p)$.

2.7.2.2 Model 2: Determination of Optimal Preventive Replacement Age

5. The objective is to determine the optimal age, t_p, at which preventive replacements should occur such that total downtime per unit time is minimized. The policy was illustrated earlier in Figure 2.26, from which it was seen that there are two possible cycles of operation.

The total downtime per unit time for preventive replacements once the item becomes of age t_p is

$D(t_p) = $ (total expected downtime/cycle)/expected cycle length

Total expected downtime/cycle = downtime due to a preventive cycle $\times$ probability of a preventive cycle + downtime due to a failure cycle $\times$ probability of a failure cycle $= T_p R(t_p) + T_f [1 - R(t_p)]$

Expected cycle length (as per Section 2.5.2) =

$$(t_p + T_p)R(t_p) + [M(t_p) + T_f][1 - R(t_p)]$$

Therefore,

$$D(t_p) = \frac{T_p R(t_p) + T_f[1 - R(t_p)]}{(t_p + T_p)R(t_p) + [M(t_p) + T_f][1 - R(t_p)]} \tag{2.11}$$

This is a model of the problem relating replacement age to total downtime.

2.7.3 NUMERICAL EXAMPLES

Let $T_f = 0.07$ week, $T_p = 0.035$ week, and $f(t) = N(5, 1)$.

2.7.3.1 Model 1: Replacement Interval

From Equation 2.10 we have

$$D(2) = \frac{H(2) \times 0.07 + 0.035}{2 + 0.035}$$

Table 2.9 gives values of $D(t_p)$ for various values of t_p. Here it can be seen that the optimal replacement interval is $t_p = 4$ weeks.

Sample calculation for $t_p = 2$:

Equation 2.10 becomes

$$D(2) = \frac{H(2) \times 0.07 + 0.035}{2 + 0.035}$$

$$H(2) = 0.00135 \text{ (from Section 2.4.4)}$$

Therefore,

$$D(2) = 0.0173 \text{ weeks}$$

TABLE 2.9
Optimal Preventive Replacement Interval:
Downtime Minimization

t_p	1	2	3	4	5	6
$D(t_p)$	0.0338	0.0173	0.0121	0.0114	0.0139	0.0156

2.7.3.2 Model 2: Replacement Age

From Equation 2.11 we have

$$D(t_p) = \frac{0.035R(t_p) + 0.07[1 - R(t_p)]}{(t_p + 0.035)R(t_p) + [M(t_p) + 0.07][1 - R(t_p)]} \tag{2.12}$$

Then Equation 2.12 becomes

$$D(t_p) = \frac{0.035R(t_p) + 0.07[1 - R(t_p)]}{(t_p + 0.035)R(t_p) + \int_{-\infty}^{t_p} tf(t)dt + 0.07[1 - R(t_p)]} \tag{2.13}$$

Inserting different values of t_p into Equation 2.13, Table 2.10 can be constructed, from which it is clear that the optimal replacement age is 4 weeks.

Sample calculation for $t_p = 3$:

Equation 2.11 becomes

$$D(3) = \frac{0.035R(3) + 0.07 \times [1 - R(3)]}{(3 + 0.035)R(3) + \int_{-\infty}^{3} tf(t)dt + 0.07 \times [1 - R(3)]}$$

From Section 2.5.3,

$$R(3) = 0.9772 \qquad 1 - R(3) = 0.0228 \qquad \int_{-\infty}^{3} tf(t)dt = 0.0600$$

Therefore,

$$D(3) = 0.0118 \text{ weeks}$$

TABLE 2.10
Optimal Preventive Replacement Age:
Downtime Minimization

t_p	1	2	3	4	5	6
$D(t_p)$	0.0338	0.0172	0.0118	0.0102	0.0113	0.0129

2.7.4 FURTHER COMMENTS

With reference to model 1, provided that the time required to affect a failure replacement is small relative to the intervals being considered for preventive replacement (e.g., 0.07 as opposed to 4), it is reasonable to use the $H(T)$ formulation of Section 2.4.3 to determine the expected number of failures in interval $(0, t_p)$. Strictly speaking, account should be taken of the fact that the time available between preventive replacements for failure to occur is reduced due to downtime that is incurred while making failure replacements.

Note also that although the replacement interval and replacement age to minimize downtime are both 4 weeks, the age-based policy gives a reduction in downtime of 10.5%, for the figures used in the example, when compared with the interval-based policy.

2.7.5 APPLICATIONS

2.7.5.1 Replacement of Sugar Refinery Cloths

The practice in a refinery was to replace certain critical components in a centrifuge only when they failed (Jardine and Kirkham, 1973). The goal was to identify an optimal change-out time for several components, including the cloth, such that machine availability was maximized. As mentioned in Section 2.3, one of the requirements necessary for preventive replacement to be worthwhile is that the probability of an item failing in service must be increasing as the item ages. In this study, when the failure statistics were analyzed (there were 229 failure intervals available for analysis) the Weibull shape parameter, β, was equal to 1.0. Thus, in this case, the downtime minimization model was not required, and the conclusion was that the best replacement policy was to continue replacing the cloths only when they failed.

In practice, one can ask the question, Why are cloths failing "strictly randomly"? In other words, the conditional probability of a new cloth failing is the same as that of an old one. As a consequence of addressing this question, perhaps a design or a change in operating procedures may be made. In the case described, the decision was taken to continue using the same type of cloth and to continue with past practice, namely, to replace the item only upon failure.

2.7.5.2 Replacement of Sugar Feeds in a Sugar Refinery

In the same study, discussed above, another component examined was the sugar feed (Jardine and Kirkham, 1973). Again, the policy in place was one of replacing the feed once it reached a defined failed state, and the goal was to establish an optimal preventive replacement time (either age or interval) such that the sugar feed downtime was minimized or, equivalently, availability was maximized. When the failure statistics were analyzed in this case, the Weibull shape parameter took the value 0.8. This was somewhat of a surprise, since it indicated that there was a high probability of sugar feed failure shortly after installation, compared to later in the life of the feed. Again, the question can be asked: Why? In this case, one possible reason is poor-quality installation process for the part, and perhaps with additional training, the installer would do an improved installation. A worthwhile consideration when dealing with the statistical analysis of data, especially if a surprise is observed, is to look behind the statistics. Perhaps there has been an error in data acquisition. In this study, the data were carefully examined and there was no reason to reject the conclusion: there is a higher risk of failure early in the life of the feed, and as it ages the risk of failure is reduced. Thus, the optimal policy is to replace the feed only when it fails.

2.8 GROUP REPLACEMENT: OPTIMAL INTERVAL BETWEEN GROUP REPLACEMENTS OF ITEMS SUBJECT TO FAILURE: THE LAMP REPLACEMENT PROBLEM

It is sometimes worthwhile to replace similar items in groups, rather than singly, since the cost of replacing an item under group replacement conditions may be lower; that is, there are economies of scale. Perhaps the classic example of this sort of situation is that of maintenance of street lamps. Bearing in mind the cost of transporting a lighting department's maintenance staff to a single street lamp failure and discounts associated with bulk purchase of lamps, it may be economically justifiable to replace all the lamps on a street rather than only the failed ones.

This particular type of problem is virtually identical to that of Section 2.4, except that here we are dealing with a group of identical items, rather than single items.

2.8.1 STATEMENT OF PROBLEM

There are a large number of similar items that are subject to failure. Whenever an item fails, it is replaced by a new item — we do not assume group replacement (i.e., replacing all items at the same time) in such conditions. There is also the possibility that group replacement can be performed at fixed intervals of time. The cost of replacing an item under group replacement conditions is assumed to be less than that for failure replacement. The more frequently group replacement is performed, the less failure replacements will occur, but a balance is required between the money spent on group replacements and the reduction of failure replacements.

The model developed for this problem is based on the assumption that the replacement policy is to perform group replacements at fixed intervals of time, with failure replacements being performed as necessary. We wish to determine the optimal interval between the group replacements in order to minimize the total expected cost of replacement per unit time.

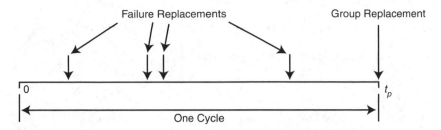

FIGURE 2.28 Group replacement.

2.8.2 CONSTRUCTION OF MODEL

1. C_g is the cost of replacing one item under conditions of group replacement.
2. C_f is the cost of a failure replacement.
3. $f(t)$ is the probability density function of the failure times of the items.
4. N is the total number of items in the group.
5. The replacement policy is to perform group replacement at constant intervals of length t_p, with failure replacements performed as many times as required in interval $(0, t_p)$. The policy is illustrated in Figure 2.28.
6. The objective is to determine the optimal interval between group replacements to minimize the total expected replacement cost per unit time.

The total expected replacement cost per unit time for group replacement at time t_p, denoted as $C(t_p)$, is

$$C(t_p) = \frac{\text{Total expected cost in interval } (0, t_p)}{\text{Interval length}}$$

Total expected cost in interval $(0, t_p)$ = cost of group replacement at time t_p + expected cost of failure replacements in interval $(0, t_p)$ = $NC_g + NH(t_p)C_f$, where $H(t_p)$ is the expected number of times one item fails in interval $(0, t_p)$. The method of determining $H(t_p)$ is given in Section 2.4.3. Therefore,

$$C(t_p) = \frac{NC_g + NH(t_p)C_f}{t_p} \tag{2.14}$$

This is a model of the group replacement problem relating replacement interval t_p to total cost.

2.8.3 NUMERICAL EXAMPLE

Using the data given in the example of Section 2.4.4, namely:

1. Cost of a failure replacement = $10
2. Cost of a replacement under group replacement conditions = $5
3. $f(t) = N(5, 1)$

and assuming that there are 100 items in the group, Table 2.11 can be constructed. It gives values of $C(t_p)$, the total replacement cost per unit time, for various values of t_p, the group replacement interval, from which it is seen that the optimal interval between group replacements is 4 weeks.

TABLE 2.11

t_p	1	2	3	4	5	6
$C(t_p)$	500	251	174	165	200	224

Sample calculation for $t_p = 2$ weeks:

Equation 2.14 becomes

$$C(2) = [100 \times 5 + 100 \times H(2) \times 10]/2$$

$$H(2) = 0.00135 \quad \text{(from example in Section 2.4.4)}$$

Therefore,

$$C(2) = [500 + 1.4]/2 = \$251$$

2.8.4 FURTHER COMMENTS

As stated prior to the problem statement of this section, the optimal group replacement interval for the above example is identical to the optimal preventive replacement interval for a single item, as given in Section 2.4. The minimum total replacement cost for group replacement is the same as that for a single unit, multiplied by the number of items in the group.

2.8.5 AN APPLICATION: OPTIMAL REPLACEMENT INTERVAL
FOR A GROUP OF 40 VALVES IN A COMPRESSOR

A group of 40 valves are presently replaced every 9000 hours on a compressor in the oil and gas industry. Examination of the company maintenance database indicates that in these 9000-hour intervals there is occasionally a valve failure, and when this occurs the defective valve is replaced.

From this data source it is possible to estimate the failure distribution of a valve, in that operating environment, as being Weibull with parameters: shape

(β) = 2, location (γ) = 3600 hours, characteristic life (η) = 138,118 hours, and mean time to failure (μ) = 126,000 hours.

The cost associated with compressor failure due to a valve problem was estimated at $94, 024.00, and for preventive replacement of the group of 40 valves it was $24,256.00.

Using the constant-interval model, Equation 2.3, the optimal change-out time for the valve is identified as 84,000 hours with an associated cost of $0.66/hour, compared to the cost under the current policy of $2.65/hour. Thus, there is a cost reduction of 76%.

However, given the limited data that were available for analysis, one may not immediately jump to adopt the replacement interval of 84,000 hours. The analysis revealed that the total cost curve was quite flat around the optimal replacement interval, and furthermore, by doubling the current replacement interval to 18,000 hours, a very substantial savings could be expected compared to the present practice, since the cost per hour would be reduced to $1.41/hour, realizing a savings of 47%.

2.9 FURTHER REPLACEMENT MODELS

Three further classes of component replacement models are outlined for the interested reader. In each case, the level of mathematics used is of a slightly higher order than that of the models developed in this book.

2.9.1 MULTISTAGE REPLACEMENT

A multistage replacement strategy may be relevant in the situation where there is a group of similar items that can be divided into subgroups dependent on the cost of replacing an item upon its failure.

For example, some items may be more expensive to replace than others due to failure in a key position, having expensive repercussions.

A two-stage replacement strategy is examined in a paper by Bartholomew (1963). The problem examined is one where there are N similar items divided into two groups, N_1 and N_2, and the costs of replacement of an item in these groups are C_1 and C_2, respectively. The two-stage replacement strategy is illustrated in Figure 2.29.

In Figure 2.29 it is assumed that the cost of replacement in stage 1 is greater than that in stage 2. In this case, all failures that occur in stage 2 are replaced by operating items from stage 1. Vacancies that occur in stage 1, whether caused by failure or transfer of operating items to stage 2, are replaced by new items. Although this strategy does not reduce the overall steady-state failure rate of the system, it does decrease it in stage 1 (where replacement cost is high) and increase it in stage 2 (where replacement cost is low). In Bartholomew's paper, the conditions are derived for two-stage replacement to be preferable to simple replacement (i.e., replacing any failure directly with a new item).

A possible application of such a strategy relates to the replacement of tires on certain classes of mobile equipment. For example, if a failure occurs in a rear tire

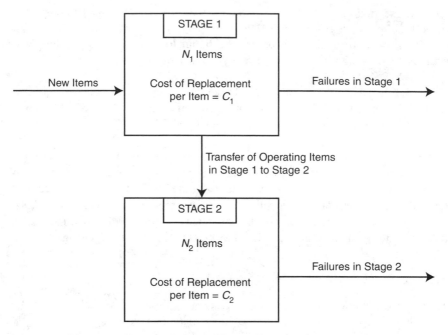

FIGURE 2.29 Two-stage replacement.

on a trailer and it is to be replaced, then it is replaced by a tire from one of the front wheels of the tractor (prime mover), and the new tire is placed on the tractor wheel.

The two-stage strategy is extended in a paper by Naik and Nair (1965) to cater for the possibility of defining several stages in a system, each stage being defined by its replacement cost. An application of the two-stage strategy to resource planning is given by Robinson (1974).

2.9.2 Optional Policies

Frequently an item ceases to operate, not because of its own failure, but because there is a production stoppage for some reason. When this happens, the maintenance specialist may have to decide whether to take advantage of the downtime opportunity to perform a preventive replacement.

Woodman (1967) discussed this class of problem and constructed a model to cover optional policies (so called because the decision on whether to take advantage of the downtime opportunity is at the option of the maintenance specialist). Basically, the model takes account of the costs of failure replacement, cost of replacement during downtime, the failure distribution of the equipment subject to replacement, and the frequency with which replacement opportunities occur. Solution of the model results in control limits being determined, which enable the specialist to determine whether to take advantage of the opportunity, depending on the age of the equipment. This policy is illustrated in Figure 2.30. If an equipment failure occurs, it is replaced. If a replacement opportunity occurs

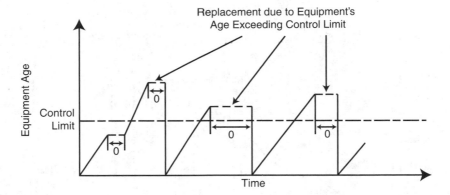

FIGURE 2.30 Optional replacement policy, 0 = replacement opportunities.

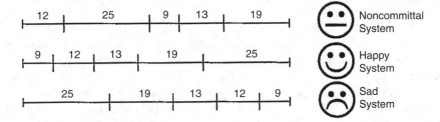

FIGURE 2.31 Repairable system maintenance.

and the equipment's age exceeds the control limit, a preventive replacement is made; otherwise, the equipment is left during the opportunity and allowed to continue operating. Kaspi and Shabtay (2003) present models using the optional replacement modeling approach for machine tools in a manufacturing setting.

2.9.3 REPAIRABLE SYSTEMS

Up until now in the chapter, it has been assumed that renewal of the item occurred at the time of the maintenance action. If this is not acceptable, then we need models applicable to repairable systems. A classic book addressing such problems is that of Ascher and Feingold (1984), where the concept of noncommittal, happy, and sad systems is introduced. Figure 2.31 illustrates these system descriptions using the five sets of bearing failure data that were first introduced in Section 2.5.5. Before proceeding to use the interval and age models presented, it is necessary that the failures are what are termed identical and independently distributed (iid); namely, the failure distribution of each new item is identical to the previous one, and each failure time is independent of the previous one. To check that this is the case, a trend test can be made on the chronologically ordered failure times (see the Laplace trend test described in Section A2.12 of Appendix 2). Figure 2.31 illustrates that the noncommittal times have no clear trend, whereas

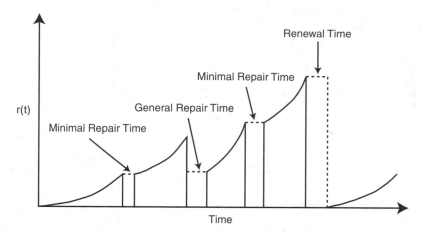

FIGURE 2.32 Minimal and general repair.

that is not the case for the times that are presented in the rows identified as "happy" or "sad." Of course, in practice, trends may not be so easy to spot, but care should always be taken before proceeding to do a Weibull analysis of data, to ensure that there is no underlying trend of reliability growth (the happy system) or reliability degradation (the sad system) that is common with repairable systems. Lhorente et al. (2004) provide an example of a sad system. Since the Ascher and Feingold book was published, many research ideas on how best to handle the optimization of maintenance decisions associated with repairable systems have been published. In this literature the terms *minimal* and *general repair* are frequently used. Figure 2.32 illustrates these two terms. Here it is seen that a minimal repair can be thought of as a very minor maintenance action (such as replacing a snapped fan belt on an automobile) that returns the equipment/system to the same state of health that it was in just before the minor maintenance action. A general repair improves the system state while a renewal completely returns the equipment to the statistically as-good-as-new condition.

The concept of virtual age has been introduced in order to model repairable system maintenance problems. Malik (1978) and Kijima (1989) presented the concept of virtual age when modeling repairable systems. Jiang et al. (2001) suggested a repair limit using the virtual age concept when deciding what maintenance action should take place at the time of a maintenance fault. The approach is illustrated in Figure 2.33. Here it can be seen that the real age (run-time) of an item is on the Y-axis at the top of the figure and virtual age is on the X-axis. Thus at the first maintenance intervention, when the equipment is of age X_1, the equipment becomes of age V_1 after the maintenance action. Once the equipment has operated for a further period X_2 to bring its running age to $X_1 + X_2$, another maintenance action occurs that brings the equipment's virtual age to V_2.

The key question to be addressed is: When there is the need for a maintenance intervention, what action should be taken? Minimal repair, general repair (the

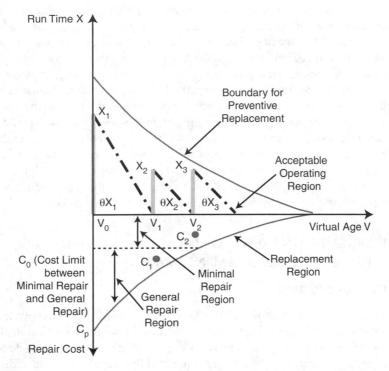

FIGURE 2.33 Optimizing minimal and general repair decisions.

more money spent on a general repair, the closer the equipment is brought to the as-new condition), or complete renewal. To address this question, we can see the repair limit concept in Figure 2.33 through the cost on the Y-axis on the bottom half of the diagram.

If the cost estimate for maintenance is between zero and the limit C_0, then a minimal repair is made. If the cost is between C_0 and the cost boundary, then a general repair is made.

In the figure, the cost associated with the first maintenance intervention was C_1, so a general repair instead of a complete renewal should be made, and consequently, the equipment health was improved when compared to its condition prior to repair commencing.

If the cost exceeds the boundary curve, then it implies that a complete renewal should take place. Similarly, if the running time of the equipment reaches the boundary for preventive replacement, then a complete renewal is to take place.

In a study on repair vs. replacement of transformers (Kallis, 2003) it was established that a new transformer could be purchased for $150,000. Key components experiencing failures were primary and secondary windings, each installed on a laminated iron core, internal insulating mediums, and the main tank and bushings. Others included the cooling system, off-circuit tap changer, under-load tap changer, current and potential transformers, and mechanism cabinets.

Optimization of repairable systems requires identification of the degree of improvement in a component's performance subsequent to repair. For windings, it was concluded that the degree of repair was 80%. Thus, changing the core and windings of the transformer reduced the age of the transformer by 80%.

Thus, if a 20-year-old transformer has its core and windings replaced, the virtual age of the transformer would be $(20 - 0.8 \times 20) = 4$ years. Similar degrees of repair were investigated for other transformer components. Knowing the cost of the maintenance action and comparing it with the cost of a new transformer, an intelligent decision can be made about the repair vs. replacement alternative.

Research publications dealing with the development of models that can be used for the optimization of maintenance decisions for repairable equipment that cannot be treated as an item that is always renewed at the maintenance action are provided by Lugtigheid et al. (2004, 2005) and Nelson (2003).

2.10 SPARE PARTS PROVISIONING: PREVENTIVE REPLACEMENT SPARES

2.10.1 INTRODUCTION

In Chapter 1, the concept of modeling was introduced through an example dealing with establishing the optimal economic order quantity for an item. That model is appropriate for fast-moving consumable spare parts where there is a steady demand.

If preventive maintenance is being conducted on a regular basis according to either the constant-interval or age-based replacement models (Sections 2.4 and 2.5, respectively), then a spare part is required for each preventive replacement, but in addition, spare parts are required for any failure replacements. The goal of this section is to present a model that can be used to forecast the expected number of spares required over a specified period, such as a year, for a given preventive replacement policy.

2.10.2 CONSTRUCTION OF MODEL

t_p is the preventive replacement time (either interval or age).
$f(t)$ is the probability density function of the item's failure times.
T is the planning horizon, typically 1 year.
$EN(T, t_p)$ is the expected number of spare parts required over the planning horizon, T, when preventive replacement occurs at time t_p.

2.10.2.1 The Constant-Interval Model

$EN(T, t_p)$ = number of preventive replacements in interval $(0, T)$
        + number of failure replacements in interval $(0, T)$
    = $T/t_p + H(t_p)(T/t_p)$

where $H(t_p)$ is defined in Section 2.4.

2.10.2.2 The Age-Based Preventive Replacement Model

$EN\ (T,\ t_p)$ = number of preventive replacements in interval $(0, T)$
+ number of failure replacements in interval $(0, T)$

In this case, the approach to take is to calculate the expected time to replacement (either preventive or failure) and divide this time into the planning horizon, T. This gives

$$EN(T, t_p) = \frac{T}{t_p \times R(t_p) + M(t_p) \times [1 - R(t_p)]}$$

where development of the denominator of the above equation is provided in Section 2.5.2.

2.10.3 NUMERICAL EXAMPLE

2.10.3.1 Constant-Interval Policy

Using the same data as in Section 2.4.4, namely, $C_p = \$5$, $C_f = \$10$, failures occur according to a normal distribution with mean = 5 weeks, standard deviation = 1 week, and the optimal preventive replacement interval is 4 weeks.

Assuming the planning horizon is 12 months (52 weeks), then the expected number of replacements will be

$$EN(52, 4) = 52/4 + 0.16\ (52/4) = 15.08 \text{ per year}$$

If there were 40 similar components in service in a plant, then the expected number of replacements per year would be 603.20; thus, 604 spares would be needed for the fleet over the year.

2.10.3.2 Age-Based Policy

Using the same data as in Section 2.5.3, namely, $C_p = \$5$, $C_f = \$10$, failures occur according to a normal distribution with mean = 5 weeks, standard deviation = 1 week, and the optimal preventive replacement age is 4 weeks.

Again, assuming the planning horizon is 12 months (52 weeks), then the expected number of replacements will be

$$EN(T, t_p) = \frac{T}{t_p \times R(t_p) + M(t_p) \times [1 - R(t_p)]}$$

$$EN\ (52, 4) = 52/(4 \times 0.84 + 3.17 \times 0.16) = 52/3.87 = 13.44 \text{ per year}$$

Once more, if there were 40 similar components in a fleet, the expected number of replacements per year would be 537.6, so 538 spares would be required.

2.10.4 FURTHER COMMENTS

Once the demand is forecast, there is the issue of acquiring the expected number of replacement parts. There is a large body of literature dealing with the area of inventory control, for example, Tersine (1988), where there are models available to assist in establishing an optimal acquisition policy, including the possibility of taking advantage of quantity discounts.

2.10.5 AN APPLICATION: CYLINDER HEAD REPLACEMENT — CONSTANT-INTERVAL POLICY

A cylinder head for an engine costs $1946, and the policy employed is to replace the eight-cylinder heads in an engine as a group at age 9000 hours, plus failure replacement as necessary during the 9000-hour cycle. In the plant there were 86 similar engines in service. Thus, over a 12-month period there is total component utilization of $8 \times 86 \times 8760 = 6,026,880$ hours worth of work.

Estimating the failure distribution of a cylinder head, and taking the cost consequence of a failure replacement as 10 times that of a preventive replacement, it was estimated that with the constant-interval replacement policy, the expected number of spare cylinder heads required per year to service the entire fleet was 849 (576 due to preventive replacement and 273 due to failure replacement).

2.11 SPARE PARTS PROVISIONING: INSURANCE SPARES

2.11.1 INTRODUCTION

A critical issue in spares management is to establish an appropriate level for insurance (emergency) spares that can be brought into service if a current long-life and highly reliable component fails. Such components would include transformers in an electrical utility or electric motors in a conveyor system. To maintain a highly reliable service, a few spare units may be kept in stock. The question to be addressed in this section is: How many critical spares should be stocked?

To answer the question, it is necessary to specify if the spare part is one that is scrapped after failure (a nonrepairable spare) or if it can be repaired and renewed after its failure and put back into stock (a repairable spare). And finally, it is necessary to understand the goal. In this section four criteria will be considered for establishing the optimal number of both nonrepairable and repairable spares. They are:

1. *Instantaneous reliability* — This is the probability that a spare is available at any given moment in time. In some literature this is known as availability of stock, fill rate, or point availability in the long run.
2. *Interval reliability* — This is the probability of not running out of stock at any moment over a specified period, such as 1 year.
3. *Cost minimization* — This takes into account costs associated with purchasing and stocking spares, and the cost of running out of a spare part.
4. *Availability* — This is the percentage of nondowntime (uptime) of a system/unit where the downtime is due to shortage of spare parts.

The detailed mathematical models behind the following analyses are provided in Louit et al. (2005).

2.11.2 CLASSES OF COMPONENTS

2.11.2.1 Nonrepairable Components

With nonrepairable components, when a component fails or has been preventively removed, it is immediately replaced by one from the stock (the replacement time is assumed to be negligible), and the replaced component is not repaired (i.e., it is discarded; see Figure 2.34). It is assumed that the demand for spares follows a Poisson process, which, for emergency parts demand, has found wide application. Several references describe models based on this principle (see, e.g., Birolini, 1999).

To introduce the mathematics behind the optimization of spare parts requirements, consider a group or fleet of m independent components to be in use for an interval of length T, with the mean time to failure of one component equal to

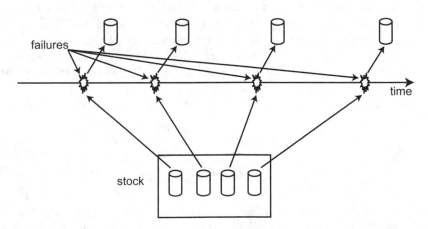

FIGURE 2.34 Nonrepairable spares.

μ and the standard deviation σ. Let $N(T, m)$ be the total number of failures in interval $[0, T]$ and $S(k, m)$ the time until the kth failure. Then the probability of having less than k failures in $[0, T]$ is equal to the probability that the time until the kth failure is greater than T, that is,

$$\Pr(N(T, m) < k) = \Pr(S(k, m) > T) \tag{2.15}$$

2.11.2.2   Normal Distribution Approach

$S(k, m)$ is asymptotically normally distributed with mean $\mu k/m$ and variance $\sigma^2 k/m^2$, for large k (Cox, 1962). That is,

$$\Pr(N(T, m) \le k) = \Pr(S(k, m) \ge T) = 1 - \Phi\left(\left(T - \frac{\mu k}{m}\right)\frac{m}{\sigma\sqrt{k}}\right) \tag{2.16}$$

where $\Phi(\cdot)$ is the cumulative standard normal distribution. In this way, it is possible to calculate the required amount of spares given a certain desired reliability of stock, p. From the equation, $\Pr(S(k, m) > T) = p$, k can be calculated as

$$k = \left(\frac{Z_p\sigma}{2\mu} + \sqrt{\left(\frac{Z_p\sigma}{2\mu}\right)^2 + \frac{Tm}{\mu}}\right)^2 \tag{2.17}$$

where Z_p is obtained from a standard cumulative normal distribution table.

2.11.2.3   Poisson Distribution Approach

The normal distribution approach described above is valid only when T is large in comparison with μ/m. The Poisson distribution also can be used. This approximation is also independent of the underlying failure distribution, and is valid for a relatively small number of components, as the superposition of component failure times converges rapidly to a Poisson process (Cox, 1962). If the underlying failure distributions are exponential, the number of failures $N(T, m)$ follows exactly the Poisson process, for any number of components, m. For a Poisson process,

$$\Pr(N(T, m) = i) = \frac{a^i}{i!}e^{-a} \tag{2.18}$$

where a is the expected number of failures in the interval $[0, T]$. For one component, the expected number of failures in $[0, T]$ is T/μ; thus for m components, $a = mT/\mu$. Now it is possible to calculate k, for which

$$\Pr(N(T,\,m) \le k) = \Pr(S(k+1,m) \ge T) = \sum_{i=0}^{k} \frac{a^i}{i!} e^{-a} \ge p \qquad (2.19)$$

The obtained value of k will be the minimum stock level that ensures a reliability p (probability of not having a stock-out; that is, there is no demand when there is no spare in stock). Note: Equation 2.18 assumes that a is not a very large number.

2.11.2.4 Repairable Components

The basic ideas associated with identifying optimal stock-holding policies for repairable components will be presented through the following example.

A group of m independent components have been in use for a time interval of length T, and now one of the components is sent to the repair shop after failure. After being repaired, the component is sent back to stock (Figure 2.35). Let s spare components be originally placed in stock, so they can instantly replace the failed components. It is also assumed that the repair is perfect; that is, the repaired component is returned to the as-new state. We will only consider the situation when there is no limit on the number of repairs that can be performed simultaneously (unlimited repair capacity). An extension to the limited repair capacity problem is discussed by Barlow and Proschan (1965, chap. 5). We are interested in determining the initial number of spares that should be kept in stock, in order to limit the risk of running out of spares. Two situations are considered:

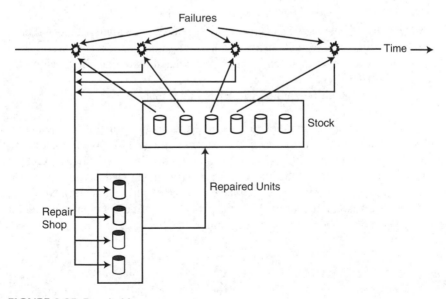

FIGURE 2.35 Repairable spares.

- *Instantaneous or point reliability* — Spares are available on demand (we must not run out of spares at any given moment).
- *Interval reliability* — Spares are available at all moments during a given interval of time (we must not run out of spares at any time during a specified interval, e.g., for 12 months). This situation is obviously more demanding than the instantaneous reliability case.

A brief summary of the modeling behind the instantaneous reliability calculation follows: The point reliability approach means that we determine the number of spare components in such a way that at any given moment in time the probability of running out of spares is less than the required reliability p. Let the average time to failure be μ for each of the m components in use. Then for each component the average rate of failures (i.e., arrivals in the repair facility) equals $1/\mu$, and for m components it equals m/μ. Let the average time to repair be μ_R, and thus the average repair rate for i components is i/μ_R. Let the number of components on repair at time t be $M(t)$. We have to find the probability of not running out of spares at t, that is, $\Pr(M(t) \le s)$.

Analogous to the Poisson approach for nonrepairable components, the probability of having less than or equal to s components on repair, for large m, can be calculated using the Poisson distribution:

$$\Pr(M(t) \le s) = \sum_{i=0}^{s} \frac{a^i}{i!} e^{-a} \qquad (2.20)$$

where $a = (\mu_R m)/\mu$ is the expected number of failures arriving during one repair. Then a represents the average number of spares required to cover failures during one repair. Now the stock level, s, can be calculated as the smallest value of s such that $\Pr(M(t) \le s) \ge p$.

Equation 2.20 can be applied when t is large, that is, in the steady state, if m is large, p small, and a not large.

A brief summary of the modeling behind the interval reliability calculation follows: Let the number of spares in the system be s, with i units in repair at the beginning of the interval $[0, t]$. Let $p_{ij}^s(t)$ be the probability of having exactly j units on repair at the end of the interval $[0, t]$ and *not* having a delay in production because of a spares shortage (we may have no spares, but no demand). Note that i and j are less than or equal to s. Then the probability of no delay during interval $[0, t]$, given i units on repair at the beginning, is

$$p_i^s(t) = \sum_{j=0}^{s} p_{ij}^s(t) \qquad (2.21)$$

So, we need to calculate the matrix $P^s(t) = \left[p_{ij}^s(t) \right]$.

This can be done using the transition rates for the states of a Markov process representing the number of units on repair at moment t. For example, the rate of transition from state i to state $j = (i + 1)$, where $i \leq s$ is simply the rate at which a new failure occurs, that is, (m/μ). (Note: Since m components are in operation and any one of them can fail at moment t.) If Q is the matrix of transition rates for the Markov process and $Q(s)$ is that same matrix truncated at s (using only the first $(s + 1)$ rows and columns), then we have

$$P^s(t) = \exp\big(tQ(s)\big). \tag{2.22}$$

Matrix exponentiation is not discussed here; see, for example, Bhat and Miller (2002) for details.

A summation over the rows of $P^s(t)$ gives the reliabilities for the interval $[0, t]$ for each initial number of units in repair, i.

The required number of spares, s, for a given interval reliability, p, can be obtained from Equation 2.21 by setting $P_i^s(t) = p$. The calculation is numerically intensive and requires computer programming (see Section 2.11.4).

2.11.3 COST MODEL

Shortages of spares may lead to extended downtime that can have important cost implications for the company. On the other hand, larger stocks imply higher inventory holding costs. Models incorporating acquisition and holding costs for spares, and cost of downtime due to stock-out are provided in Louit et al. (2005).

In many applications, it is likely that the cost of a component will vary considerably depending on the conditions of procurement. Normally, the cost is lower if there is no urgency for receiving the item, whereas it is very likely that premiums apply in emergency situations.

With the incorporation of cost considerations, optimization can be performed for any of the four criteria described in Section 2.11.1.

2.11.4 FURTHER COMMENTS

With use of the models for spare parts demand presented in this chapter, prototype software was developed that allows for the determination of optimal stocking policies according to the optimization of several performance measures. This tool has been found to be of great value for companies operating in industries characterized by the intensive use of physical assets.

Some extensions for the prototype software include the incorporation of restrictions on the capacity of the repair facility, the specification of an arrival distribution for failed parts, and calculation of the expected duration of plant downtime in the case of repairable components.

2.11.5 An Application: Electric Motors

A total of 62 electric motors were used simultaneously in conveyor belts in a mining operation, and the company was interested in determining the optimal number of motors to stock. The motors are expensive, and even purchasing one extra motor was considered a significant investment (see Wong et al., 1997). The answer to this question is not unique, but depends on the objective of the company, that is, what is to be optimized in selecting the stock size. With the problem specifications presented in Table 2.12, the prototype software known as Spares Management Software (SMS) was used to do the following exercise.

Results are obtained for assumptions of nonrepairable and repairable spares, as shown in Table 2.13 and Table 2.14. Note that in the nonrepairable components' situation, two cases are considered: (1) strictly random (constant arrival rate; thus, there is a Poisson arrival process for failures) and (2) not strictly random (arrival rate not constant, but failure distribution given by a mean and standard deviation). Also, in the repairable components' case, unlimited repair capacity is assumed, which is realistic due to the expected number of motors that would be on simultaneous repair.

Some comments on the results: In the nonrepairable case it will be noticed that while the required reliability was 95%, the associated reliability in each case was higher, 95.61% in one case and 97.63% in the other. The reason for this is that the resulting stock level has to be an integer, and for the random failure case, if the stock level was set at 47, rather than 48, then the 95% required reliability

---

TABLE 2.12
Example Parameters for the System of Conveyors

Parameter	Value
Number of components (motors) in operation	62
Mean time to failure (μ)	3000 days (8 years)
Planning horizon (T)	1825 days (5 years)
	Note: The planning horizon could be much shorter and may, for example, be close to the mean repair time of a component since one may want to ensure with a high probability not running out of stock while a component is being repaired.
Mean time to repair (μ_R)	80 days
Cost of one spare component (regular procurement)	$15,000
Cost of one spare component (emergency procurement)	$75,000
Value of unused spare after 5 years	$10,000
Holding cost for one spare	$4.11 per day (10% of value of part per annum)
Cost of conveyor's downtime for a single motor	$1000 per day

TABLE 2.13
Solution for Nonrepairable Spares

Case and Optimization Criteria	Optimal Stock Level[a]	Associated Reliability
(i) Random failures:		
95% reliability required	48	95.61%
Cost minimization	47	94.02%
(ii) Not strictly random failures (= 1000 days):		
95% reliability required	42	97.63%
Cost minimization	41	93.80%

[a] Total number of spares required during the planning horizon of 5 years. Note: There is then the need to decide how best to acquire the spares over the 5-year planning horizon.

TABLE 2.14
Solution for Repairable Spares

Case and Optimization Criteria	Optimal Stock Level	Associated Availability
Random failures:		
95% interval reliability required	7	Not calculated
95% instant reliability required	4	Not calculated
95% availability required	0	97.40%
Cost minimization	6	99.99%

would not have been achieved. At least 48 items are required, and at 48 the associated reliability is 95.61%.

In the repairable spares example it will be noticed that to achieve a 95% availability, the stock level required is zero. The reason for this outcome is due to the repair time for a failed motor, 80 days, being very short in comparison to the mean time to failure of a motor, 8 years.

2.12 SOLVING THE CONSTANT-INTERVAL AND AGE-BASED MODELS GRAPHICALLY: USE OF GLASSER'S GRAPHS

2.12.1 INTRODUCTION

Glasser (1969) wrote a paper in which he presented two graphs that can be used to quickly identify the optimal preventive replacement time (interval, called block by Glasser, or age) of an item provided the item can be assumed to have failure times that can be described by a Weibull distribution, and that the objective is to minimize total cost. The graphs are provided in Figure 2.36 and Figure 2.37.

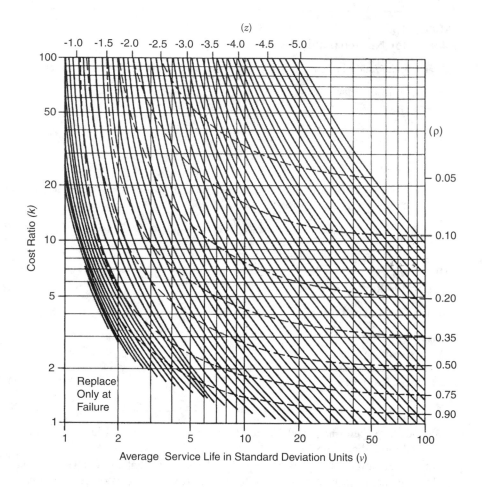

FIGURE 2.36 Glasser's graph: optimal policies under block replacement — Weibull distribution. (Reprinted with permission of ASQ.)

Furthermore, Glasser's graphs provide an indication of the economic benefits of changing the item at the optimal replacement time, as opposed to a run-to-failure policy. Since the Weibull is so flexible (see Appendix 2), Glasser's graphs are very helpful to quickly identify the best change-out time of an item and establish the savings that can result from implementation of the policy. If the economic savings are worthwhile, then it may be advantageous to use one of the classical replacement models to determine more precisely the economic replacement time. A further benefit of using the model is that a very clear picture is provided of the form of the total cost curve around the region in which the optimal solution lies. As mentioned earlier (Section 2.2.4), this knowledge can be very valuable to management in making a final decision. (Use of Glasser's graphs can

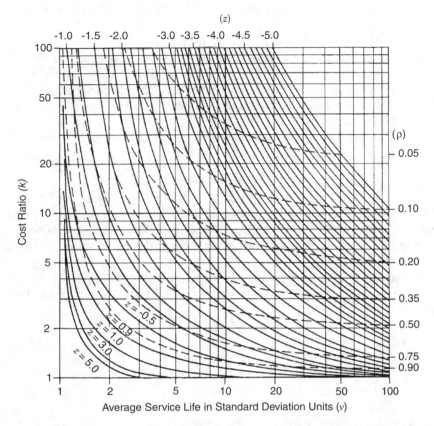

FIGURE 2.37 Glasser's graph: optimal policies under age replacement — Weibull distribution. (Reprinted with permission of ASQ.)

only provide two cost points on the total cost curve: given the cost of a failure or preventive replacement, the cost associated with the optimal replacement time can be calculated, as can the cost associated with a replace-only-on-failure [R-o-o-F] policy.)

2.12.2 USING GLASSER'S GRAPHS

There are three assumptions, none of which are very restrictive:

1. The failure distribution is Weibull and its mean (μ) and standard deviation (σ) are known.
2. The ratio of the cost of a failure replacement to that of a preventive replacement is known, namely, C_f/C_p.
3. The objective is total cost minimization.

There are four straightforward steps to the solution procedure:

1. Obtain the cost ratio denoted by k $(= C_f/C_p)$ and mark it on the Y-axis of the graph.
2. Obtain the ratio $\upsilon = \mu/\sigma$ and mark it on the X-axis.
3. Obtain the Z value from the graph. The Z value is obtained by determining the intersection of a horizontal line from k and a vertical line from υ, then following a solid line to the Z scale that is on the X-axis at the top of the graph sheet. Interpolation between two solid lines may be required.
4. The optimal preventive replacement time (interval or age) is obtained from solving the equation $t_p = \mu + Z\sigma$.

2.12.3 Example

Using the same data used in Sections 2.4.4 (interval replacement) and 2.5.3 (age replacement), namely, $C_p = \$5$, $C_f = \$10$, and the failure distribution is normal, that is, $f(t) \sim N(5, 1)$, determine the optimal preventive replacement interval to minimize total cost.

Note that while the distribution is specified as normal, the Weibull distribution can serve as a good approximation to the normal distribution if $\beta = 3.5$. Thus, we can proceed to use Glasser's graph.

Solution:

1. $k = Cf/Cp = 10/5 = 2$
2. $v = \mu/\sigma = 5/1 = 5$
3. $Z = -1.2$
4. $t_p = \mu + Z\sigma = 5 + (-1.2) \times 1 = 3.8$ weeks

This equals 4.0 weeks if rounded up, which is the same answer as obtained in Section 2.4.4 (note that in Section 2.4.4 the problem was solved numerically with only integer values being used).

2.12.4 Calculation of the Savings

It will be noticed that on the right-hand side of Glasser's graphs there is a ρ scale on the Y-axis. The ρ value identifies the cost of the optimal policy as a decimal fraction of the cost associated with following a replace-only-on-failure (R-o-o-F) policy.

The ρ value is again obtained from the intersection of a horizontal line from k and a vertical line from υ, and then following to the right a dotted line, interpolating if required. In this example ρ is estimated from the graph as 0.85. Therefore, the optimal policy of $t_p = 3.8$ weeks costs 85% of a R-o-o-F policy.

Since it is known that $C_f = \$10$, then

$$\text{Cost of a R-o-o-F policy} = C_f/\mu = 10/5 = \$2/\text{week}$$

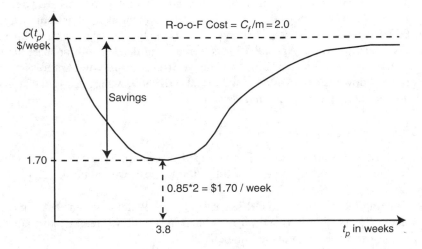

FIGURE 2.38 Cost function: Glasser's graph.

The optimal policy cost is $0.85 \times 2 = \$1.70$ per week, and the savings as a percentage is 15%. This is illustrated in Figure 2.38.

2.13 SOLVING THE CONSTANT-INTERVAL AND AGE-BASED MODELS USING THE OREST SOFTWARE

2.13.1 INTRODUCTION

Rather than solve the mathematical models for component preventive replacement interval or age, from first principles we saw in the previous section how a graphical solution can be used. A disadvantage of graphical solutions is lack of precision compared to using a mathematical model. Software that has the models programmed in provides a very easy way to solve the models, and also provides a high level of accuracy. One such package is OREST (Optimal Replacement of Equipment in the Short Term), which has been developed based on material in this book.

OREST will take item failure and suspension times (for a definition and description of suspensions, see Section A2.7 of Appendix 2) and will fit a Weibull distribution to the data. To estimate the parameters of the Weibull distribution, OREST uses the approach of median rank regression analysis. It is not the purpose of this book to address in detail various criteria that can be used to estimate distribution parameters, but the most common from an engineer's perspective would be to use regression analysis (this is the approach presented in Appendix 2), whereas from a statistician's viewpoint, maximum likelihood would be the preferred criterion. In practice, if the data set being analyzed is large, then there will be little difference in the parameter estimates.

Once the Weibull parameters are estimated, OREST will then provide the option of establishing the optimal preventive replacement interval or optimal age. OREST has a number of other features, such as analyzing for possible trends in data (see Section A2.12, Appendix 2) and forecasting the demand for spare parts. The interested reader is referred to the Web site http://www.crcpress.com/ e_products/downloads/download.asp?cat_no=DK9669, where the educational version of OREST can be downloaded free.

2.13.2 USING OREST

We will use the bearing failure data provided in Section 2.5.5, namely, the five failure times, ordered from the shortest to the longest, which are 9, 12, 13, 19, and 25 weeks.

Entering these values into OREST provides the Weibull parameter estimates β = 2.67 and η = 17.57, based on regression analysis. A screen capture of the parameter estimation is provided in Table 2.15.

If required, OREST also fits a three-parameter Weibull to the data and the result can be compared to the standard two-parameter distribution.

Using the values of β = 2.67, η = 17.57, we get the cost function depicted in Figure 2.39 and the age-based preventive replacement report shown in Table 2.16, from which it is seen that the optimal preventive replacement age is 6.39 weeks. While this preventive replacement age might seem small compared to the shortest observed failure time of 9 weeks, the Weibull analysis has assumed that

TABLE 2.15
OREST Weibull Parameter Estimates

OREST - Weibull Parameter Estimation

Current Component
Bearing

Data Summary

Data Type	Ungrouped	Number of Failures	5
Age Unit	Week	Number of Suspensions	0

Estimated Parameters

Shape	2.67	Mean Life	15.62	Median Life	15.31
Scale	17.57	Standard Deviation	6.3	B10 Life	7.56
Location	0	Characteristic Life	17.57		

Goodness of Fit - Kolmogorov-Smirnov Test

Kolmogorov-Smirnov Test Statistic	0.24	p-Value	0.9

Test Result: The hypothesis that the Weibull fits the data is NOT rejected at 5% significance level.

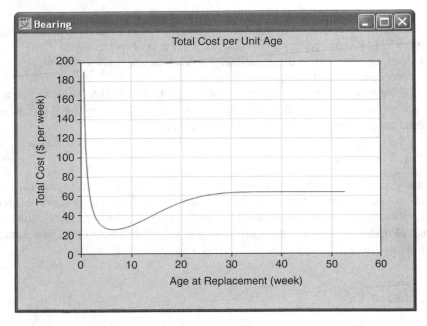

FIGURE 2.39 OREST: cost optimization curve.

TABLE 2.16
OREST Age-Based Preventive Replacement Report

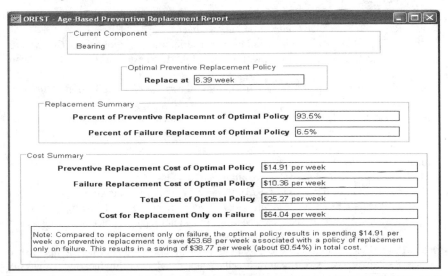

in practice a bearing failure could occur shortly after installation, and so a two-parameter Weibull has been used. Furthermore, the consequence of failure is quite severe ($1000) compared to the cost of a preventive replacement ($100). If required, one could use a three-parameter Weibull and so preclude the possibility of the occurrence of a short failure time.

It again should be stressed that software such as OREST enables many sensitivity checks to be undertaken, so that one can establish a robust recommendation on the optimal change-out time for an item.

2.13.3 FURTHER COMMENTS

This section has just dipped very briefly into one software package that can be used to optimize the preventive replacement times for a component. Others include RelCode, developed by Hastings initially in 1976, but regularly updated, and Weibull++ (www.weibull.com). Hastings (2004) presents a case study that illustrates the use of RelCode.

REFERENCES

Ascher H. and Feingold, H. (1984). *Repairable Systems Reliability.* New York: Marcel Dekker.

Barlow, R. and Proschan, F. (1965). *Mathematical Theory of Reliability.* New York: Wiley. Reprinted in 1996 by the Society of Industrial and Applied Mathematics, Philadelphia, in their book series SIAM Classics in Applied Mathematics.

Bartholomew, D.J. (1963). Two-stage replacement strategies. *Operational Research Quarterly,* 14.

Bhat, V.N. and Miller, G.L. (2002). *Elements of Applied Stochastic Processes,* 3rd ed. New York: Wiley.

Birolini, A. (1999). *Reliability Engineering,* 3rd ed. Berlin: Springer.

Cox, D.R. (1962). *Renewal Theory.* London: Methuen.

Davidson, D. (1970). An overhaul policy for deteriorating equipment. In *Operational Research in Maintenance*, A.K.S. Jardine, Ed. Manchester University Press, chap. 5, pp. 72–99.

Glasser, G.J. (1969). Planned replacement: some theory and its application. *Journal of Quality Technology,* 1.

Hastings, N.A.J. (2004). Component reliability, replacement and cost analysis with incomplete failure data. In *Case Studies in Reliability and Maintenance*, W.R. Blischke and D.N. Prabhakar Murthy, Eds. New York: Wiley, chap. 16.

Jardine, A.K.S. (1979). Solving industrial replacement problems. In *Proceedings, Annual Reliability and Maintenance Symposium*, pp. 136–142.

Jardine, A.K.S. and Kirkham, A.J.C. (1973). Maintenance policy for sugar refinery centrifuges. In *Proceedings of the Institution of Mechanical Engineers*, Vol. 187, pp. 679–686.

Jiang, X., Makis, V., and Jardine, A.K.S. (2001). Optimal repair/replacement policy for a general repair model. *Advances in Applied Probability,* 33.

Kallis, C. (2003). Economic Life Calculation of Distribution Transformers. Unpublished B.A.Sc. thesis, Department of Mechanical and Industrial Engineering, University of Toronto.

Kapur, K. and Lamberson, L. (1977). *Reliability in Engineering Design*. New York: Wiley.

Kaspi, M. and Shabtay, D. (2003). Optimization of the machining economics problem for a multistage transfer machine under failure, opportunistic and integrated replacement strategies. *International Journal of Production Research*, 41, 2229–2247.

Kijima, M. (1989). Some results for repairable systems with general repair. *Journal of Applied Probability*, 26, 89–102.

Lhorente, B., Lugtigheid, D., Knights, P.F., and Santana, A. (2004). A model for optimal armature maintenance in electric haul truck wheel motors: a case study. *Reliability Engineering and System Safety*, 84, 209–218.

Louit, D., Banjevic, D., and Jardine, A.K.S. (2005). Optimization of Spare Parts Inventories Composed of Repairable or Non-repairable Parts. Paper presented at Proceedings, ICOMS, Hobart, Australia.

Lugtigheid, D., Banjevic, D., and Jardine, A.K.S. (2004). Modeling repairable systems reliability with explanatory variables and repair and maintenance actions. *IMA Journal of Management Mathematics*, 15, 89–110.

Lugtigheid, D., Banjevic, D., and Jardine, A.K.S. (2005). Component Repairs: When to Perform and What to Do. Paper presented at Annual Reliability and Maintenance Symposium.

Malik, M.A.K. (1978). Reliable preventive maintenance scheduling. *AIIE Transactions*, R28, 331–332.

Naik, M.D. and Nair, K.P.K. (1965). Multi-stage replacement strategies. *Operations Research*, 13.

Nelson, W.B. (2003). Recurrent Events Data Analysis for Product Repairs, Disease Recurrences, and Applications, ASA-SIAM.

Robinson, D. (1974). 2-stage replacement strategies and their application to manpower planning. *Management Science Series B — Application*, 2, 199–208.

Smith, W.L. (1955). Regenerative stochastic processes. In *Proceedings of the Royal Statistical Society A*, Vol. 232.

TEAM (Technical and Engineering Aids to Management). (1976). *Technical and Engineering Aids to Management*, Vol. 4, Winter.

Tersine, R.J. (1988). *Principles of Inventory and Materials Management*. Amsterdam: Elsevier.

Wong, J.Y.F., Chung, D.W.C., Ngai, B.M.T., Banjevic, D., and Jardine, A.K.S. (1997). Evaluation of spares requirements using statistical and probability analysis techniques. *Transactions of Mechanical Engineering, IEAust.*, 22, 77–84.

Woodman, R.C. (1967). Replacement policies for components that deteriorate. *Operational Research Quarterly*, 18.

PROBLEMS

To following problems are to be solved using the mathematical models.

1. The hydraulic pump used for tipping the box in a garbage truck becomes less efficient with usage. This results in less productivity in terms of value of materials moved per month. The average cost of replacing a pump is $1200. The trend in productivity is as follows:

TABLE 2.17
Component Failure Frequency Data

Thousands of Kilometers to Failure	0–10	10–20	20–30	30–40
Expected Frequency	250	250	250	250

Month 1 since new	$10,000 worth of material is moved
Month 2 since new	$9700 worth of material is moved
Month 3 since new	$9400 worth of material is moved
Month 4 since new	$8900 worth of material is moved

What is the optimal replacement time for the pump to minimize the total cost of replacement and lost production?

2. A car-rental company has kept records on a particular vehicle component. Although failure of the component is random, it is a function of vehicle use. Data on 1000 failures have been collected and analyzed, and through using a χ^2 goodness-of-fit test, the conclusion is reached that the time to failure of the component can be described adequately by a uniform distribution. Table 2.17 gives the distribution of the expected frequencies of the 1000 failures.

The total cost of a preventive replacement of the component is $100. A failure results in a penalty cost being incurred, and in total, the cost of a failure replacement is $200. It is reasonable to assume that the times taken to effect either a preventive or failure replacement are negligible.

The car rental company wishes to consider implementing a preventive replacement policy. The particular policy it is interested in is frequently termed an age-based policy; that is, it is one where preventive replacement occurs only when a component has reached a specific age, say t_p; otherwise, a failure replacement is made.

Considering values of t_p = 10,000, 20,000, 30,000, and 40,000 km, which one gives the smallest expected total cost per 1000 km? Clearly explain the derivation of any model you use and your line of reasoning in reaching a conclusion.

3. Truck battery failures have been analyzed, and through using a statistical goodness-of-fit test, it has been concluded that the battery failures can be assumed to follow the expected frequency distribution shown in Figure 2.40.

It is known that the average cost associated with a battery failure is $900 (including spoilage of goods), whereas a preventive replacement can be undertaken at a total cost of $300. Given that $C_f > C_p$, the truck fleet operator wishes to consider implementing an age-based replacement policy using the following model:

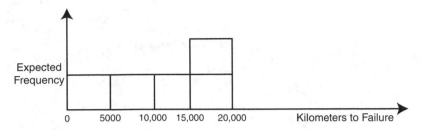

FIGURE 2.40 Battery failure data.

$$C(t_p) = \frac{C_p \times R(t_p) + C_f \times [1 - R(t_p)]}{t_p \times R(t_p) + M(t_p) \times [1 - R(t_p)]}$$

What is the optimal preventive replacement age? Consider values of t_p = 5000, 10,000, 15,000, and 20,000 km.

4. It is known that failure of the pump results in the vehicle being out of service for approximately 3 days, whereas the replacement of a pump on a preventive basis takes an average of half a day. The pump failures are assumed to follow a distribution with a p.d.f. $f(t) = \begin{cases} 0.2, 0 \leq t \leq 2 \\ 0.1, 2 < t \leq 8 \end{cases}$

Using the following model for minimization of total downtime, determine the optimal preventive replacement age (only consider values of t_p = 2, 4, 6, and 8 months):

$$D(t_p) = \frac{T_p \times R(t_p) + T_f \times [1 - R(t_p)]}{t_p \times R(t_p) + M(t_p) \times [1 - R(t_p)]}$$

5. Water pump failures from a fleet of transit vehicles have been analyzed, and a χ^2 goodness-of-fit test allows the hypothesis to be accepted that pumps fail according to a uniform distribution in the range 0 to 20,000 km.

It is known that the average unavailability associated with water pump failures is 9 days (because of limited personnel), whereas a preventive replacement can be undertaken with only 3 days unavailability of a transit vehicle. Given that $D_f > D_p$, the maintenance officer wishes to implement an age-based preventive replacement policy using the following model:

$$D(t_p) = \frac{D_p \times R(t_p) + D_f \times [1 - R(t_p)]}{t_p \times R(t_p) + M(t_p) \times [1 - R(t_p)]}$$

What is the optimal preventive replacement age? Only consider values of t_p = 5000, 10,000, 15,000, and 20,000 km and show your method of solution.

TABLE 2.18
Radiator Failure Data

Class (K = 10³ miles)	F(t)
0K < 5K	0.0250
5K < 10K	0.0500
10K < 15K	0.0625
15K < 20K	0.1726
20K < 25K	0.2917
25K < 30K	0.3231
30K < 35K	0.3566
35K < 40K	0.3933
40K < 45K	0.4949
45K < 50K	0.5813
50K < 55K	0.5943
55K < 60K	
60K < 65K	0.6050
65K < 70K	
70K < 75K	
75K < 80K	0.9500

TABLE 2.19
Component Failure Times (hours)

115	80	150	200	130	170	100

The following problems are to be solved using Glasser's graphs (several of the problems first require that a Weibull analysis be undertaken for the failure data).

6. The cumulative probability data of Table 2.18 relate to radiator failure. Given that a failure replacement is five times as costly as a preventive replacement, what is the optimal preventive replacement interval to minimize total cost per 1000 miles?

7. A component gave the times to failure of Table 2.19.
   If a preventive replacement costs $50 and a failure replacement $500, and the objective is to minimize total cost/unit time, what is:

   a. The optimal age-based preventive replacement policy?
   b. The optimal constant-interval preventive replacement policy?

   In each case, state the cost of the optimal policy as a percentage of a replace-only-on-failure policy.

8. The Moose Fleet operator has kept records on a particular vehicle component. Although failure of the component is random, it is a function of vehicle use. Data have been collected and analyzed, and through using a χ^2 goodness-of-fit test, the conclusion is reached that the time to failure of the component can be described adequately by a Weibull distribution having a mean of 20,000 km and standard deviation of 1000 km.

   The total cost of a preventive replacement of the component is $100. A failure results in a penalty cost being incurred, and in total, the cost of a failure replacement is $200. It is reasonable to assume that the time taken to effect either a preventive or failure replacement is negligible.

   The Moose Fleet operator wishes to consider implementing a preventive replacement policy. The particular policy he is interested in is frequently termed an age-based policy; that is, it is one where preventive replacement only occurs when a component has reached a specific age, say t_p; otherwise, a failure replacement is made.

   What is the optimal preventive replacement age and what percentage cost savingd does it give over a replace-only-on-failure policy?

9. Bearing failure in the blower used in diesel engines in semi-tractors has been determined as occurring according to a Weibull distribution with a mean life of 150,000 km, with a standard deviation of 10,000 km. Failure in service of the bearing results in costly repairs, and in total, a failure replacement is 10 times as expensive as a preventive replacement.

   a. Determine the optimal preventive replacement interval (or block policy) to minimize total cost per kilometer.
   b. What is the expected cost savings associated with your optimal policy over a replace-only-on-failure policy?
   c. Given that the cost of a failure replacement is $2000, what is the cost per kilometer associated with your optimal policy?

10. Irrespective of the age of a component, the replacement policy to be adopted is one where preventive replacements occur at fixed intervals of time and failure replacements take place when necessary.

   a. Making appropriate assumptions, construct a model that could be used to identify a replacement policy such that total cost per unit time is minimized. Very clearly explain each step in the construction of your model.
   b. Given the following data, solve the model that you construct in (a):
      The labor and material cost associated with a preventive or failure replacement is $50.
      The value of production losses associated with a preventive replacement is $50, while for a failure replacement it is $100.
      Failure distribution is normal, with a mean of 200 hours and standard deviation of 10 hours.

Also indicate the approximate cost of your optimal policy as a percentage of a failure replacement policy.

11. A sugar refinery centrifuge is a complex machine composed of many parts that are subject to sudden failure. A particular component, the plough-setting blade, is considered to be a candidate for preventive replacement, and you are required to determine an optimal replacement policy. The policy you are to consider is sometimes termed a block replacement or constant intervals, say t_p, with failure replacements taking place when necessary. Determine the optimal policy so that total cost per unit time is minimized given the following data:

   a. The labor and material cost associated with a preventive or failure replacement is $200.
   b. The value of production losses associated with a preventive replacement is $100, while that for a failure replacement is $700.
   c. The failure distribution of the setting blade can be described adequately by a Weibull distribution with mean = 150 hours and standard deviation = 15 hours.

   Also indicate the approximate cost of your optimal policy as a percentage of a replace-only-on-failure policy.

The following problems are to be solved using OREST software.
Note: The educational version of OREST restricts the number of observations that can be analyzed to six (failures plus suspensions). Also, it requires that the cost consequence of a failure replacement be $1000, and for preventive replacement it is $100. All the following problems satisfy these constraints.

12. Heavy-duty bearings in a steel forging plant have failed after the number of weeks of operation provided in Table 2.20.

   a. Use OREST to estimate the following Weibull parameters: β, η, and mean life.
   b. The cost of preventive replacement is $100 and the cost of failure replacement is $1000. Determine the optimal replacement policy.

TABLE 2.20
Bearing Failure Times

Age at Failure (weeks)

8
12
14
16
24
One unfailed at 24 weeks

TABLE 2.21
Fan Belt Failures

Truck 1	Truck 2
51,220	45,380
68,060	103,510

---

   c. The forge is cleaned and serviced once per week. Preventive replacement of the bearing can be carried out as part of this maintenance activity. At what age should the bearing be replaced, given that, in addition to direct-cost considerations, there is a safety argument for minimizing failure.

     Support your conclusions by giving the cost and the proportions of failure replacements for some alternative policies.

   d. There are two similar forging plants and each works for 50 weeks per year. Estimate the number of replacement parts required per year if the policy is preventive replacement at age 6 weeks. How many failure replacements will occur per year (steady-state average) under this policy?

13. Records from two heavy-duty dump trucks show that fan belt failures occurred at the odometer readings (kilometers, from new) listed in Table 2.21.

    At present, the odometer readings are 115,680 km for truck 1 and 132,720 km for truck 2.

   a. Prepare reliability data in a form suitable for analysis by OREST.
   b. Determine the following Weibull parameters: shape parameter β, scale parameter η, and mean life.
   c. What type of failure pattern is indicated (early life, random, wear-out?)
   d. Create the Weibull probability plot. Do you observe any trends, besides those given by the parameters?
   e. The preventive replacement cost is $100, and the failure replacement cost is $1000. Determine the optimal preventive replacement age, the cost under this policy, and the savings under this policy when compared with a policy of replacement only on failure.
   f. Preventive replacement can only be carried out at odometer readings that are multiples of 5000 km. Select an appropriate preventive replacement age. What is the cost ($/km) for this policy? How does this compare with the cost for the optimal policy?
   g. If the company has a fleet of 30 similar dump trucks, each of which averages 50,000 km per year, estimate the number of replacement fan belts that will be needed per year, under an appropriate replacement policy. Note: This can be calculated using the material of Section 2.10.

TABLE 2.22
Centrifuge Cloth Failures

Age in Hours	Failure Replacement	Preventive Replacement
0–9.99	14	0
10–19.99	5	0
20–29.99	2	4
30–39.99	1	8

TABLE 2.23
Bus Engine Failure Data

Age Range (km)	Failure Replacement	Survivors
0–49,999	2	35
50,000–99,999	8	27
100,000–149,999	33	118

h. If 30 dump trucks average 50,000 km per year, estimate the number of in-service fan belt failures that will occur, given that the policy is to replace fan belts on a preventive basis at 20,000 km. Note: This can be calculated using the material of Section 2.10.

14. The cloth filter on a sugar centrifuge is currently replaced on a preventive basis if a suitable opportunity occurs and the cloth has been in use for at least 20 hours. The cloth is also replaced on failure.

   The centrifuge cloth failure data provided in Table 2.22 are available for 10-hour time intervals of cloth life.

   a. Use OREST to analyze the failures and estimate the following parameters: shape parameter β, scale parameter η, and mean life.
   b. Is the current policy correct? What policy do you recommend?
   c. The company has three centrifuges, each of which runs an average of 400 hours per month. Estimate the number of replacement cloths required per month under the existing and recommended replacement policies.

15. A metropolitan transport company operates a fleet of similar buses. Engine failures necessitating replacement have occurred in the kilometer ranges shown in Table 2.23, which also shows the number of engines currently running in each age range.

TABLE 2.24
Alternator Warranty Data

Age Range (km)	Failure Replacement	Survivors
0–4999	1	48
5000–9999	3	123
10,000–14,999	2	104

a. Use OREST to estimate the following parameters: shape parameter β, scale parameter η, and mean life.
b. From the Weibull probability plot, estimate the 90% reliable life.
c. From the Weibull failure rate plot, estimate the age at which the instantaneous failure rate first exceeds one failure per 100,000 km.
d. The cost for failure replacement is known to be roughly 10 times the cost for preventive replacement. Use the optimal age replacement policy to answer the following:
  i. Determine the optimal replacement policy.
  ii. Under the optimal replacement policy, how many replacements will occur on average per 1,000,000 vehicle km, and what proportion of these will be failure replacements? Note: This can be calculated using the material of Section 2.10.
e. If the cost of preventive replacement is $1000 and the cost of failure replacement is $10,000, what will be the cost per 50,000 km of the following policies?
  i. Replacement only on failure
  ii. Preventive replacement as determined in (d)(i)
16. A new type of car has recently been released and is subject to warranty. An analysis of warranty claims shows several alternator failures, although as a proportion of the whole population, the number is quite small. You are involved in the analysis of warranty claims. The engineering manager asks you whether the statistical data are suitable for Weibull analysis, and if so, what conclusions can be drawn. The available data are provided in Table 2.24.

What type of failure is indicated?

a. Does this suggest faulty manufacture or a design defect?
b. The design department states that its brief called for 90% confidence of 90% reliability over a 20,000-km warranty period. Do the data available indicate that this criterion has been met?

Note: This can be calculated using the material of Section A2.4.

3 Inspection Decisions

> All business proceeds on beliefs, or judgments of probabilities, and not on certainties.

Charles W. Eliot

3.1 INTRODUCTION

The goal of this chapter is to present models that can be used to determine optimal inspection schedules, that is, the points in time at which the inspection action should take place.

The basic purpose behind an inspection is to determine the state of equipment. Once indicators, such as bearing wear, gauge readings, and quality of the product, which are used to describe the state, have been specified, and the inspection made to determine the values of these indicators, then some further maintenance action may be taken, depending on the state identified. When the inspection should take place ought to be influenced by the costs of the inspection (which will be related to the indicators used to describe the state of the equipment) and the benefits of the inspection, such as detection and correction of minor defects before major breakdown occurs.

The primary goal addressed in this chapter is that of making a system more reliable through inspection. In the context of the framework of the decision areas addressed in this book, we are addressing column 2 of the framework, as highlighted in Figure 3.1. One special class of problem also addressed in this chapter is that of ensuring with a high probability that equipment used in emergency circumstances, often called protective devices, is available to come into service if the need arises.

Three classes of inspection problems are addressed in this chapter:

1. Inspection frequencies: For equipment that is in continuous operation and subject to breakdown.
2. Inspection intervals: For equipment used only in emergency conditions (failure-finding intervals).
3. Condition monitoring of equipment: Optimizing condition-based maintenance decisions.

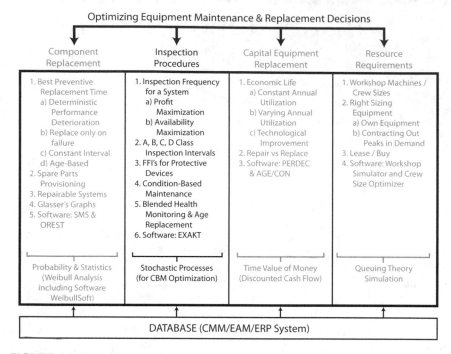

FIGURE 3.1 Inspection decisions.

3.2 OPTIMAL INSPECTION FREQUENCY: MAXIMIZATION OF PROFIT

3.2.1 STATEMENT OF PROBLEM

Equipment breaks down from time to time, requiring materials and tradespeople to repair it. Also, while the equipment is being repaired, there is a loss in production output. In order to reduce the number of breakdowns, we can periodically inspect the equipment and rectify any minor defects that may otherwise eventually cause complete breakdown. These inspections cost money in terms of materials, wages, and loss of production due to scheduled downtime.

What we want to determine is an inspection policy that will give us the correct balance between the number of inspections and the resulting output, such that the profit per unit time from the equipment is maximized over a long period.

Such a system is depicted in Figure 3.2, where it is seen that the complex system can fail for many reasons, such as caused by component 1, or component 2, and so on. And each of these causes of equipment failure could have its own independent failure distribution. Of course, it does not need to be a physical component that causes the equipment to cease functioning; it could well be a software problem that is the cause (mode) of equipment failure. Clearly, as the frequency or intensity of inspections increases, there is an expectation that the frequency of equipment/system failures will be reduced. The challenge is to identify the optimal frequency/intensity.

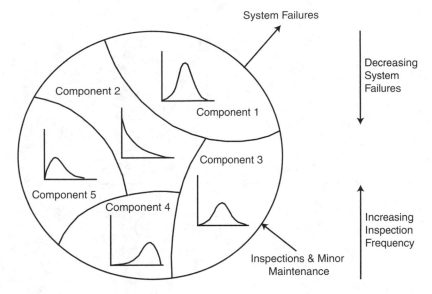

FIGURE 3.2 System failures.

3.2.2 CONSTRUCTION OF MODEL

1. Equipment failures occur according to the negative exponential distribution with mean time to failure (MTTF) = $1/\lambda$, where λ is the mean arrival rate of failures. (For example, if the MTTF = 0.5 year, then the mean number of failures per year = $1/0.5 = 2$, i.e., $\lambda = 2$.)
    Note: It is not unreasonable to make this exponential assumption for complex equipment (Drenick, 1960).
2. Repair times are negative exponentially distributed with mean time, $1/\mu$.
3. The inspection policy is to perform n inspections per unit time. Inspection times are negative exponentially distributed with mean time $1/i$.
4. The value of the output in an uninterrupted unit of time has a profit value V (e.g., selling price less material cost less production cost). That is, V is the profit value if there are no downtime losses.
5. The average cost of inspection per uninterrupted unit of time is I.
6. The average cost of repairs per uninterrupted unit of time is R.
    Note that I and R are the costs that would be incurred if inspection or repair lasted the whole unit of time. The actual costs incurred per unit time will be proportions of I and R.
7. The breakdown rate of the equipment, λ, is a function of n, the frequency of inspection. That is, the breakdowns can be influenced by the number of inspections, therefore $\lambda \equiv \lambda(n)$, as illustrated in Figure 3.3.
    In Figure 3.3, $\lambda(0)$ is the breakdown rate if no inspection is made, and $\lambda(1)$ is the breakdown rate if one inspection is made. Thus, from the figure it can be seen that the effect of performing inspections is to increase the mean time to failure of the equipment.

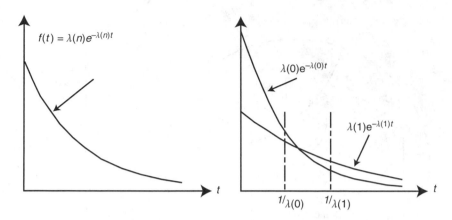

FIGURE 3.3 Breakdown rate as a function of inspection frequency.

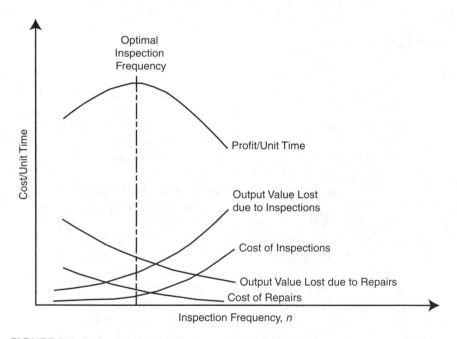

FIGURE 3.4 Optimal inspection frequency to maximize profit.

8. The objective is to choose n in order to maximize the expected profit per unit time from operating the equipment. The basic conflicts are illustrated in Figure 3.4.

The profit per unit time from operating the equipment will be a function of the number of inspections. Therefore, denoting profit per unit time by $P(n)$,

$P(n)$ = value of output per uninterrupted unit of time
  − output value lost due to repairs per unit time
  − output value lost due to inspections per unit time
  − cost of repairs per unit time
  − cost of inspections per unit time

Output value lost due to repairs per unit time
  = value of output per uninterrupted unit of time
  × number of repairs per unit time
  × mean time to repair
  = $V\lambda(n)/\mu$

Note that $\lambda(n)/\mu$ is the proportion of unit time that a job spends being repaired.

Output value lost due to inspections per unit time
  = value of output per uninterrupted unit of time
  × number of inspections per unit time
  × mean time to inspect
  = Vn/i

Cost of repairs per unit time
  = cost of repairs per uninterrupted unit of time
  × number of repairs per unit time
  × mean time to repair
  = $R\lambda(n)/\mu$

Cost of inspections per unit time
  = cost of inspection per uninterrupted unit of time
  × number of inspections per unit time
  × mean time to inspect

  = In/i

$$P(n) = V - \frac{V\lambda(n)}{\mu} - \frac{Vn}{i} - \frac{R\lambda(n)}{\mu} - \frac{In}{i} \tag{3.1}$$

This is a model of the problem relating inspection frequency n to profit $P(n)$. To get an approximate answer, we assume $P(n)$ to be a continuous function of n, so

$$\frac{dP(n)}{dn} = -\frac{V\lambda'(n)}{\mu} - \frac{V}{i} - \frac{R\lambda'(n)}{\mu} - \frac{I}{i}$$

where $\lambda'(n) = \dfrac{d}{dn}\lambda(n)$. Therefore,

$$0 = \frac{\lambda'(n)}{\mu}(V+R) + \frac{1}{i}(V+I)$$

$$\lambda'(n) = -\frac{\mu}{i}\left(\frac{V+I}{V+R}\right) \tag{3.2}$$

If values of μ, i, V, R, I, and the form of $\lambda(n)$ are known, the optimal frequency to maximize profit per unit time is that value of n that is the solution of Equation 3.2.

3.2.3 NUMERICAL EXAMPLE

Assume that the breakdown rate varies inversely with the number of inspections, that is, $\lambda(n) = k/n$, which gives

$$\lambda'(n) = -k/n^2 \tag{3.3}$$

Note that the constant k can be interpreted as the arrival rate of breakdowns per unit time when one inspection is made per unit time.

Substituting Equation 3.3 into Equation 3.2, the optimal value of n is

$$n = \sqrt{\frac{ik}{\mu}\left(\frac{V+R}{V+I}\right)}$$

Let:

Average number of breakdowns per month, k, when one inspection is made per month = 3

Mean time to perform a repair $1/\mu$ = 24 hours = 0.033 month

Mean time to perform an inspection $1/i$ = 8 hours = 0.011 month

Value of output per uninterrupted month V = \$30,000

Cost of repair per uninterrupted month R = \$250

Cost of inspection per uninterrupted month I = \$125

$$n = \sqrt{\frac{3 \times 0.033}{0.011}\left(\frac{30000+250}{30000+125}\right)} = 3.006$$

Thus, the optimal number of inspections per month in order to maximum profit is 3.

Substitution of $n = 3$ into Equation 3.1 will of course give the expected profit per unit time resulting from this policy. Insertion of other values of n into Equation

3.1 will give the expected profit resulting from other inspection policies. Comparisons can be made with the savings of the optimal policy over other possibilities, and over the policy currently adopted for the equipment.

3.2.4 FURTHER COMMENTS

The goal was to develop a model that related inspection frequency to profit. The way in which the model was developed is such that had the goal been to establish the optimal inspection frequency to minimize total cost, then the same result would have been obtained. It should be noted, however, that not all solutions that aim at maximizing profit result in the same answer as those that aim at minimizing cost.

The most important point to note from this problem is that it is concerned with identifying the best level of preventive maintenance (in the form of inspections and consequent minor overhauls and replacements) when the failure rate of equipment is constant. With complex equipment the failure distribution is negative exponential, although some individual components of the equipment may exhibit wear-out characteristics. The effect of the inspections is that certain potential component failures will be identified that, if left neglected, would cause failure of the complete equipment. If they are attended to, components will still cause equipment failure, and the overall failure distribution of the equipment will in most cases remain negative exponential, but at a reduced rate of failure. Figure 3.5 illustrates that the effect of performing inspections is to reduce the level of the failure rate. In effect, the problem is to identify the best failure rate.

The assumption was implied in the inspection problem that the depth (or level) of inspection was specified (e.g., do online monitoring of specified signals or open up equipment and take measurements x, y, and z; compare with standards; renew or do not renew components). There may also be the problem of identifying the best level of inspection. The greater the depth, the greater the inspection cost, but there is perhaps a greater chance that potential failures will be detected. In this case, a balance is required between the costs of the various possible levels of inspection and the resulting benefits, such as reduced downtime due to failures. This class of problem was originally presented in White et al. (1969).

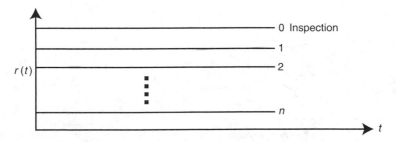

FIGURE 3.5 Effect on system failure rate of inspection frequency.

Before leaving this problem, it is worth noting that in practice, relating the failure rate of the equipment to the frequency of inspection may be difficult. One method of attack is experimentation with one's own equipment. Alternatively, if several companies have the same type of equipment doing much the same type of work, collaboration among the companies may result in determining how the failure rate is influenced by various inspection policies. Yet another approach would be to simulate different inspection frequencies. To do this would require a detailed understanding of the various ways in which the equipment could fail, and knowing the duration of the many symptoms that would indicate impending failure. Christer (1973) initially described this duration as lapse time, then later Christer and Waller (1984) described the duration as the delay time. Moubray (1991) terms it the *P-F* interval.

3.3 OPTIMAL INSPECTION FREQUENCY: MINIMIZATION OF DOWNTIME

3.3.1 STATEMENT OF PROBLEM

The problem of this section is analogous to that of Section 3.2.1: equipment breaks down from time to time, and to reduce the breakdowns, inspections and consequent minor modifications can be made. The decision now, however, is to determine the inspection policy that minimizes the total downtime per unit time incurred due to breakdowns and inspections, rather than to determine the policy that maximizes profit per unit time. Figure 3.6 illustrates the problem.

3.3.2 CONSTRUCTION OF MODEL

1. $f(t)$, $\lambda(n)$, n, $1/\mu$, and $1/i$ are defined in Section 3.2.2.
2. The objective is to choose n to minimize total downtime per unit time.

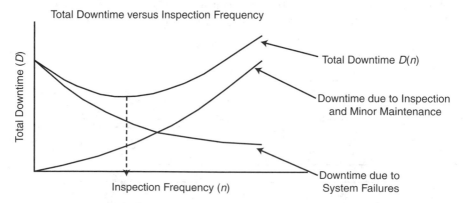

FIGURE 3.6 Optimal inspection frequency: minimizing downtime.

The total downtime per unit time will be a function of the inspection frequency, n, denoted as $D(n)$. Therefore,

$$D(n) = \text{downtime incurred due to repairs per unit time}$$
$$+ \text{downtime incurred due to inspection per unit time}$$

$$= \frac{\lambda(n)}{\mu} + \frac{n}{i} \tag{3.4}$$

Equation 3.4 is a model of the problem relating inspection frequency n to total downtime $D(n)$.

3.3.3 NUMERICAL EXAMPLE

Using the data of the example of Section 3.2.3 and assuming $D(n)$ to be a continuous function of n,

$$D(n) = \frac{\lambda(n)}{\mu} + \frac{n}{i} \text{ (from Equation 3.4)}$$

Now $\lambda'(n) = -k/n^2$, and therefore,

$$D'(n) = -\frac{k}{n^2\mu} + \frac{1}{i} = 0$$

Thus,

$$n = \sqrt{\frac{ki}{\mu}} = \sqrt{\frac{3 \times 0.033}{0.011}} = 3 \text{ inspections/month}$$

3.3.4 FURTHER COMMENTS

It will be noted that the optimal inspection frequency to minimize downtime for the above example is the same as when it is required to maximize profit (Section 3.2.3). This is not always the case. The models used to determine the frequencies are different (Equations 3.1 and 3.4), and it is only because of the specific cost figures used in the previous example that the solutions are identical for both examples.

Note also that if the problem of this section had been to determine the optimal inspection frequency to maximize availability, then this would be equivalent to minimizing downtime (since availability/unit time = 1 – downtime/unit time). Thus, in the above example, where the optimal value of $n = 3$, the minimum total downtime per month is (from Equation 3.4)

$$D(3) = \frac{3}{3} \times 0.033 + 3 \times 0.011 = 0.066 \text{ month}$$

Maximum availability = $(1 - 0.066)$ month $\equiv 93.4\%$

3.3.5 An Application: Optimal Vehicle Fleet Inspection Schedule

Montreal transit operates one of the largest bus fleets in North America, having some 2000 buses in its fleet. Buses, like most equipment, both fixed and mobile, are often subject to a series of inspections, some at the choice of the operator, while others may be statutory. The policy in Montreal was to inspect its buses at 5000-km intervals, at which an A, B, C, or D depth of inspection took place. The policy is illustrated in Table 3.1. The question to be addressed was: What is the best inspection interval to maximize the availability of the bus fleet?

While the policy was as depicted in Table 3.1, in practice, buses sometimes were inspected before a 5000-km interval had elapsed, and others were delayed. Because of that fact, it was possible to identify the relationship between the rate at which buses had defects requiring repair and different inspection intervals. In terms of the three alternatives identified in Section 3.2.4 of how to establish this relationship, the approach taken in this study could be considered experimental. While a real experiment did not occur, since different intervals in practice were being used, the conclusion can be considered to have resulted from an experiment (Jardine and Hassounah, 1990).

TABLE 3.1
Bus Inspection Policy

Kilometers (1000)	Inspection Type			
	A	B	C	D
5	X			
10		X		
15	X			
20			X	
25	X			
30		X		
35	X			
40			X	
45	X			
50		X		
55	X			
60			X	
65	X			
70		X		
75	X			
80				X
Total	8	4	3	1 $\Sigma = 16$

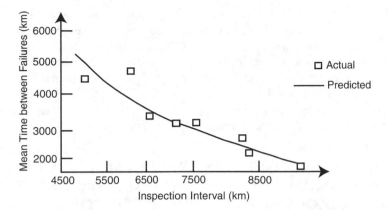

FIGURE 3.7 Mean distance to failure.

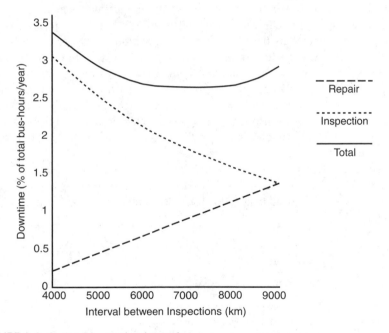

FIGURE 3.8 Optimal inspection interval.

Figure 3.7 shows the relationship between the mean time to breakdown of a bus — due to any cause — and the inspection interval. Thus, for an inspection policy of conducting inspections at multiples of 7500 km, the mean distance traveled by a bus before a defect is reported as 3000 km.

Using a slight modification to the model presented in Section 3.3.2, the total downtime curve was established, Figure 3.8, from which it is seen that minimum downtime, or maximum availability, occurs when the inspection policy is set at

8000 km. Note, however, that the curve is fairly flat within the region 5000 to 8000 km, and the final outcome was not to formally change the inspection policy from one where inspections were planned to take place at multiples of 5000 km to one where the interval would be set at 8000 km. Of course, had there been a significant benefit in increasing the interval, then that may have justified a change in policy.

Jia and Christer (2003) present an inspection interval-modeling case study that makes a comparison with the policy that would be set using the methodology of reliability-centered maintenance (RCM).

3.4 OPTIMAL INSPECTION INTERVAL TO MAXIMIZE THE AVAILABILITY OF EQUIPMENT USED IN EMERGENCY CONDITIONS, SUCH AS A PROTECTIVE DEVICE

3.4.1 STATEMENT OF PROBLEM

Equipment such as fire extinguishers and many military weapons are stored for use in an emergency. If the equipment can deteriorate while in storage, there is the risk that when it is called into use, it will not function. To reduce the probability that equipment will be inoperable when required, inspections can be made, sometimes termed proof-checking, and if equipment is found to be in a failed state, it can be repaired or replaced, thus returning it to the as-new condition. Inspection and repair or replacement take time, and the problem is to determine the best interval between inspections to maximize the proportion of time that the equipment is in the available state. A list of such items that are often called protective devices is provided in Table 3.2.

The topic of this section is to establish the optimal inspection interval for protective devices, and this interval is called the failure-finding interval (FFI). The RCM methodology addresses this form of maintenance, and Moubray (1997) has said:

TABLE 3.2
Examples of Protective Devices

Fire hydrant on city street
Standby diesel generator for runway lights
Full-face oxygen mask in aircraft cockpit
Automatic transfer switches for emergency power supply
Methane gas detector in underground coal mine
Protective relays in electrical distribution
Fire suppression system on vehicle
Hot box detector on railway car
Eye wash station in chemical plant
Refrigerant leakage detector system in chiller plant
Life raft on ship

"Failure-finding applies only to hidden or unrevealed failures. Hidden failures in turn only affect protective devices.

If RCM is correctly applied to almost any modern, complex industrial system, it is not unusual to find that up to 40% of failure modes fall into the hidden category. Furthermore, up to 80% of these failure modes require failure finding, so up to one third of the tasks generated by comprehensive, correctly applied maintenance strategy development programs are failure-finding tasks.

A more troubling finding is that at the time of writing, many existing maintenance programs provide for fewer than one third of protective devices to receive attention at all (and then at inappropriate intervals). . . .

This lack of awareness and attention means that most of the protective devices in industry — our last line of protection when things go wrong — are maintained poorly or not at all.

This situation is completely untenable."

Clearly the optimization of FFIs is an important maintenance decision topic.

3.4.2 CONSTRUCTION OF MODEL

1. $f(t)$ is the density function of the time to failure of the equipment.
2. T_i is the time required to effect an inspection. It is assumed that after the inspection, if no major faults are found requiring repair or complete equipment replacement, the equipment is in the as-new state. This may be as a result of minor modifications being made during the inspection.
3. T_r is the time required to make a repair or replacement. After the repair or replacement it is assumed that the equipment is in the as-new state.
4. The objective is to determine the interval t_i between inspections in order to maximize availability per unit time.

Figure 3.9 illustrates the two possible cycles of operation.

The availability per unit time will be a function of the inspection interval t_i. This is denoted as $A(t_i)$:

$$A(t_i) = \text{expected availability per cycle/expected cycle length}$$

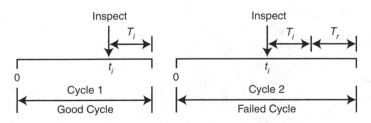

FIGURE 3.9 Maximizing availability.

The uptime in a good cycle equals t_i since no failure is detected at the inspection. If a failure is detected, then the uptime of the failed cycle can be taken as the mean time to failure of the equipment, given that inspection takes place at t_i.

Thus, the expected uptime per cycle is

$$t_i R(t_i) + \frac{\int_{-\infty}^{t_i} tf(t)dt}{1 - R(t_i)}[1 - R(t_i)]$$

$$= t_i R(t_i) + \int_{-\infty}^{t_i} tf(t)dt \quad \text{(compare with the denominator of Equation 2.8)}$$

The expected cycle length is

$$= (t_i + T_i)R(t_i) + (t_i + T_i + T_r)[1 - R(t_i)]$$

Therefore,

$$A(t_i) = \frac{t_i R(t_i) + \int_{-\infty}^{t_i} tf(t)dt}{t_i + T_i + T_r[1 - R(t_i)]} \tag{3.5}$$

This is a model of the problem relating inspection interval t_i to availability per unit time $A(t_i)$.

3.4.3 Numerical Example

1. The time to failure of equipment is normally distributed with a mean of 5 months and a standard deviation of 1 month.
2. $T_i = 0.25$ month.
3. $T_r = 0.50$ month.

Equation 3.5 becomes

$$A(t_i) = \frac{t_i R(t_i) + \int_{-\infty}^{t_i} tf(t)dt}{t_i + 0.25 + 0.50[1 - R(t_i)]}$$

TABLE 3.3
Inspection Interval vs. Availability

t_i	1	2	3	4	5	6
$A(t_i)$	0.8000	0.8905	0.9173	0.9047	0.8366	0.7371

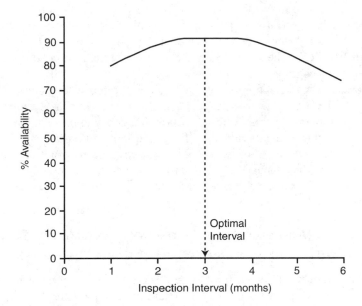

FIGURE 3.10 Optimal inspection interval.

Table 3.3 results from evaluating the right-hand side of Equation 3.5 for various values of t_i. The optimal inspection interval to maximize availability is seen to be 3 months. Figure 3.10 shows the result graphically. From a practical decision-making perspective, graphical representations are very helpful to management in making the final decision.

Sample calculation:

When $t_i = 3$ months,

$$\int_{-\infty}^{3} tf(t)dt = 0.0600$$

$$R(3) = 0.9772 \qquad \text{(see Section 2.5.3)}$$

$$1 - R(3) = 0.0228$$

Therefore, Equation 3.5 becomes

$$A(3) = \frac{3 \times 0.9772 + 0.0600}{3 + 0.25 + 0.5(0.0228)} = 0.9173$$

3.4.4 FURTHER COMMENTS

The crucial assumption in the model of this section is that equipment can be assumed to be as good as new after inspection if no repair or replacement takes place. In practice, this may be reasonable, and it will certainly be the case if the failure distribution of the equipment was exponential (since the conditional probability remains constant).

If the as-new assumption is not realistic and the failure distribution has an increasing failure rate, then rather than having inspection at constant intervals, it may be advisable to increase the inspection frequency as the equipment gets older. Such problems are discussed by Jiang and Jardine (2005).

Rather than having in place a single protective device, one can increase protection through redundancy. A discussion of various forms of redundancy (active, m-out-of-n, standby, parallel, or triple active redundancy) is presented in O'Connor (2002).

3.4.5 EXPONENTIAL FAILURE DISTRIBUTION AND NEGLIGIBLE TIME REQUIRED TO EFFECT INSPECTION AND REPAIR/REPLACEMENT

It is not unreasonable to expect protective devices to be highly reliable with the risk of failure occurring to be very low and strictly random. Also, it is not unreasonable to assume that the time required to inspect a protective device is very short (measured in minutes or hours) when compared to the optimal failure-finding interval (measured in months/years). If in the availability maximization model of Equation 3.5 we let $f(t)$ = exponential with mean time to failure (MTTF) = $1/\lambda$, $T_i = T_r = 0$, then Equation 3.5 can be reduced to

$$A(t_i) = 1 - \frac{t_i \lambda}{2}$$

Simplifying the notation, if we let

FFI = the inspection interval (t_i)
A = availability of the protective device, given an FFI
M = MTTF = $1/\lambda$

then we get the result

$$A = 1 - \left(\frac{FFI}{2M} \right) \tag{3.6}$$

3.4.6 AN APPLICATION: PRESSURE SAFETY VALVES IN AN OIL AND GAS FIELD

There are 1000 safety valves in service. The present practice is to inspect them annually. During the inspection visit 10% of the valves are found to be defective. The duration of the inspection is 1 hour. It takes an additional hour to replace each defective valve.

What is valve availability for different inspection intervals? To estimate the mean time to failure of a valve, we can use the ratio of the total testing time and the number of failures. Thus, 1000 valves have been in service for 1 year, and during that year 100 fail (10%). Therefore,

Mean time to failure is estimated from 1000/100 = 10 years (520 weeks)

Since the inspection time and replacement time are very small compared to the 12-month period (8760 hours), it is reasonable to assume these times are zero. If we further assume that the valves fail exponentially, we can estimate valve availability from Equation 3.6.

Table 3.4 provides the expected availabilities obtained from different FFIs. Thus, it is seen that at present, the practice of inspecting the valves annually provides an availability level of 95%. If an availability of 99.5% is required, then the FFI would be 5 weeks. Figure 3.11 provides a graphical representation of the relationship between availability and the failure-finding interval.

TABLE 3.4
FFIs for Pressure Safety Valve

Failure-Finding Interval (weeks)	Pressure Valve Availability (%)
1	99.9
5	99.5
10	99.0
15	98.6
21	98.1
52	95.0
104	90.0

3.5 OPTIMIZING CONDITION-BASED MAINTENANCE (CBM) DECISIONS

3.5.1 INTRODUCTION

In Chapter 2 we examined the optimization of change-out times for components subject to failure. In that chapter we estimated the probability of an item failing in service as a function of its age. A major disadvantage of the time-based

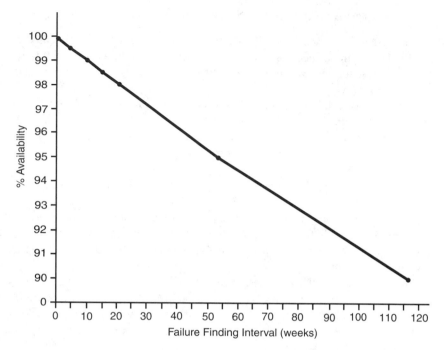

FIGURE 3.11 Availability vs. FFI.

replacement decision is that some useful life is still left in an item that has been replaced preventatively. However, taking into account the consequence of failure, it is often justified to undertake preventive replacements. On the other hand, if the item is an expensive one, such as a vehicle's transmission, rather than an inexpensive bearing, it would be worthwhile to inspect it regularly before removing it from service. Through this condition monitoring (CM) it may be possible to obtain a better understanding of the health of the item, and thus intervene with an appropriate maintenance action just before failure, and so increase the useful life of the item.

The most common form of inspection is "low tech," such as visual inspection, but for expensive items that have a long life, two common "high-tech" tools used for condition monitoring are oil analysis and vibration monitoring. Moubray (1997) in his book devotes an appendix to identifying various forms of condition monitoring, including dynamic, particle, and chemical monitoring. O'Hanlon (2003) stated that "world class companies often devote up to 50% of their entire maintenance resources to condition monitoring and the planned work that is required as a result of the findings." Clearly, condition monitoring is a key maintenance tactic in many organizations.

While much research and product development in the area of condition-based maintenance (CBM) focuses on data acquisition, such as designing tools and acquiring data, and signal processing to remove noise from the signals, the focus

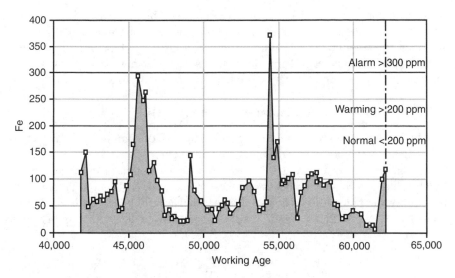

FIGURE 3.12 Classical approach to condition monitoring.

of this section of the book is to examine what might be thought of as the final step in the CBM process — optimizing the decision-making step.

Jardine (2002) provides an overview of the following procedures being used to assist organizations in making smart CBM decisions: physics of failure, trending, expert systems, neural networks, and optimization models. Possibly the most common approach to understanding the health of equipment is through plotting various measurements and comparing them to specified standards. This procedure is illustrated in Figure 3.12, where measurements of iron deposits in an oil sample are plotted on the *Y*-axis and compared to warning and alarm limits. The maintenance professional then takes remedial action if deemed appropriate. Many software vendors addressing the needs of maintenance have packages available to assist in trending CM measurements, with the goal of predicting failure.

A consequence observed when such an approach is undertaken is that the maintenance professional is often too conservative in interpreting the measurements. In work undertaken by Anderson et al. (1982), it was observed that 50% of the aircraft engines that were removed before the end of life for which they were designed, due to information obtained through sampling of engine oil, were identified by the engine manufacturer to still be in a fit state to remain on the four-engine aircraft. Christer (1999) observed the same point, where he reported that since condition monitoring of gearboxes was introduced, gearbox failures within an organization had fallen by 90%. As Christer said: "This is a notable accolade for CM." He also reported that when reconditioning "defective" gearboxes, in 50% of the cases there was no evident gearbox fault. He then concluded: "Seemingly, CM can be at the same time very effective, and rather inefficient."

Clearly there is a need to focus attention on the optimization of condition-monitoring procedures. In this section we will present an approach for estimating the hazard (conditional probability of failure) that combines the age of equipment and condition-monitoring data using a proportional hazards model (PHM). We will then examine the optimization of the CM decision by blending in with the hazard calculation the economic consequences of both preventive maintenance, including complete replacement, and equipment failure.

3.5.2 THE PROPORTIONAL HAZARDS MODEL (PHM)

A valuable statistical procedure for estimating the risk of equipment failing when it is subject to condition monitoring is the PHM (Cox, 1972). There are various forms that can be taken by a PHM, all of which combine a baseline hazard function along with a component that takes into account covariates that are used to improve the prediction of failure. The particular form used in this section is known as a Weibull PHM, which is a PHM with a Weibull baseline and is

$$h(t, Z(t)) = \frac{\beta}{\eta} \left(\frac{t}{\eta} \right)^{\beta-1} \exp \left\{ \sum_{i=1}^{m} \gamma_i z_i(t) \right\} \tag{3.7}$$

where $h(t, Z(t))$ is the (instantaneous) conditional probability of failure at time t, given the values of $z_1(t), z_2(t), ..., z_m(t)$.

Each $z_i(t)$ in Equation 3.7 represents a monitored condition data variable at the time of inspection, t, such as the parts per million of iron or the vibration amplitude at the second harmonic of shaft rotation. These condition data are called covariates. The γ_i values are the covariate parameters that along with the Z_i values indicate the degree of influence each covariate has on the hazard function.

The model consists of two parts: the first part is a baseline hazard function that takes into account the age of the equipment at time of inspection, $\frac{\beta}{\eta} \left(\frac{t}{\eta} \right)^{\beta-1}$,

and the second part, $e^{\gamma_1 z_1(t) + \gamma_2 z_2(t) + \cdots + \gamma_m z_m(t)}$, takes into account the variables that may be thought of as the key risk factors used to monitor the health of equipment, and their associated weights.

In the study by Anderson et al. (1982) the form of the hazard model for the aircraft engines was

$$h(t) = \frac{4.47}{24100} \left(\frac{t}{24100} \right)^{3.47} \exp \left(0.41 z_1(t) + 0.98 z_2(t) \right)$$

where $z_1(t)$ is iron (Fe) concentration and $z_2(t)$ is chromium (Cr) concentration in parts per million, and t is the age of the aircraft engine in flying hours at the time of inspection. Since $\beta = 4.47$, we know that the age of the aircraft engine is an influencing factor in estimating the hazard rate of the engine; $\eta = 24,100$ hours is the scale parameter of the Weibull PHM.* The values 0.41 and 0.98 are weights given to the iron and chromium measurements when calculating the hazard rate. They are estimated from the data that are analyzed and may be different for engines of different types, and may depend on their operating environment.

The procedure to estimate the values of β, η, and the weights, along with determining the condition-monitoring variables to be included in the model, is discussed in a number of books and papers, including Vlok et al. (2002) and Kalbfleisch and Prentice (2002).

Standard statistical software such as SAS and S-Plus have routines to fit a PHM — both parametric, such as the Weibull PHM, and nonparametric.

3.5.3 BLENDING HAZARD AND ECONOMICS: OPTIMIZING THE CBM DECISION

Makis and Jardine (1992) presented an approach to identify the optimal interpretation of condition-monitoring signals. The approach is illustrated graphically in Figure 3.13 and Figure 3.14.

Figure 3.13 illustrates that given a set of condition-monitoring measurements (the data plot), it is possible to convert the measurements to the equivalent hazard estimate (the risk plot). This conversion is achieved through using a PHM.

Once we have a method of monitoring an equipment's hazard value, the next question is: What should we do about it to make an optimal maintenance decision? The answer is illustrated in Figure 3.14. There it can be seen that one possibility is to ignore risk (see risk plot graph). If risk information is ignored, then the equipment will be used until it fails, and only then will it be maintained (for the time being, assume that the maintenance action is equivalent to a replacement, as is the case of some complex equipment, such as aircraft engines, where after maintenance the engines are "relifed" and have the same guarantees as a new engine). The cost associated with this decision (ignoring risk) is the cost of a failure replacement divided by the mean time to failure of the equipment. Thus, we obtain the cost of replacing only on failure, as identified in the cost plot. As the risk level (threshold) is reduced, there will be more preventive replacement actions and less failure replacements. Assuming that the cost of a failure replacement is greater than the cost of a preventive replacement, a cost function as illustrated on the cost plot will be obtained. Thus, it is possible to identify the optimal hazard level at which the equipment should be replaced: if the hazard rate is greater than the threshold value, preventive replacement should take place; otherwise, operations can continue as normal.

In the Makis and Jardine (1992) paper, it is shown that the expected average cost per unit time, $\Phi(d)$, is a function of the threshold risk level, d, and is given by

---

* Note: η in this case does not take the interpretation that 63.2% of failures occur before this time, which would be the case if the hazard was not influenced by covariates.

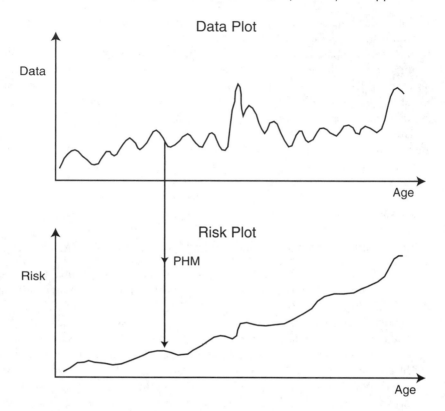

FIGURE 3.13 Calculating hazard from condition-monitoring measurements.

$$\Phi(d) = \frac{C\big(1 - Q(d)\big) + \big(C + K\big)Q(d)}{W(d)} \tag{3.8}$$

where C is the preventive replacement cost and $C + K$ the failure replacement cost. $Q(d)$ represents the probability that failure replacement will occur, at hazard level d. $W(d)$ is the expected time until replacement, either preventive or at failure.

The optimal risk, d^*, is that value that minimizes the right-hand side of Equation 3.8, and the optimal decision is then to replace the item whenever the estimated hazard, $h(t, Z(t))$, calculated on completion of the condition-monitoring inspection at t, exceeds d^*.

3.5.4 Applications

The topic of optimizing CBM decisions has been an active research thrust at the University of Toronto that has been conducted for some years in partnership with a number of companies, many of them having global operations (www.mie.utoronto.ca/cbm). As a consequence, pilot studies have been undertaken and published in

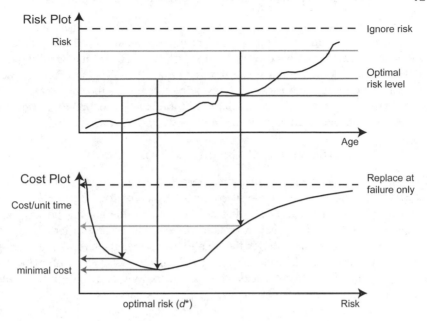

FIGURE 3.14 Establishing the optimal hazard level for preventive replacement.

the open literature. Brief summaries of three of them, each utilizing a different form of condition monitoring, are in the sections below.

3.5.4.1 Food Processing: Use of Vibration Monitoring

A company undertook regular vibration monitoring of critical shear pump bearings. At each inspection 21 measurements were provided by an accelerometer. Using the theory described in the previous section, and its embedding in software called EXAKT (see Section 3.5.6), it was established that of the 21 measurements, there were 3 key vibration measurements: velocity in the axial direction in both the first bandwidth and the second bandwidth, and velocity in the vertical direction in the first bandwidth.

In the plant, the economic consequence of a bearing failure was 9.5 times greater than when the bearing was replaced on a preventive basis. Taking account of risk as obtained from the PHM and the costs, it was clear that through following the optimization approach it was estimated that total cost could be reduced by 35% (Jardine et al., 1999).

3.5.4.2 Coal Mining: Use of Oil Analysis

Electric wheel motors on a fleet of haul trucks in an open-pit mining operation were subject to oil sampling on a regular basis. Twelve measurements resulted from each inspection. These were compared to warning and action limits in order to decide whether the wheel motor should be preventively removed. These measurements were Al, Cr, Ca, Fe, Ni, Ti, Pb, Si, Sn, Visc 40, Visc 100, and sediment.

After applying a PHM to the data set, it was identified that there were only two key risk factors: iron (Fe) and sediment measurements — oil analysis measurements that were highly correlated to the risk of the wheel motor failing due to the failure modes being monitored through oil analysis. The cost consequence of a wheel motor failure was estimated as being three times the cost of replacing it preventively, and the economic benefit of following the optimal replacement strategy was an estimated cost reduction of 22% (Jardine et al., 2001).

3.5.4.3 Transportation: Use of Visual Inspection

Traction motor ball bearings on trains were inspected at regular intervals to determine the color of the grease; it could be in one of four states: light gray, gray, light black, or black. Depending on the color of the grease and knowing the next inspection time, a decision was made to either replace or leave the ball bearings in service. As a result of building a PHM relating the hazard of a bearing failing before the next planned inspection, a decision was made to dramatically reduce the interval between checks from 3.5 years to 1 year. Before the study was undertaken, the transportation organization was suffering, on average, nine train stoppages per year. The expected number with a reduced inspection interval was estimated to be one per year. The year following the study the transportation system identified two system failures due to a ball bearing defect. The overall economic benefit was identified as a reduction in total cost of 55%. It should be mentioned that this included the cost of additional inspectors and took into account the reduction in passenger disruption. A notional cost was identified with passenger delays.

3.5.5 FURTHER COMMENTS

Additional case studies dealing with the optimization of CBM decisions in a variety of sectors using the optimization approach presented in this section are Willets et al. (2001), pulp and paper; Vlok et al. (2002), coal plant; Jardine et al. (2003), nuclear plant refueling; Lin et al. (2003), military land armored vehicle; Monnot et al. (2004), construction industry backhoes; Jefferis et al. (2004), marine diesel engines; and Chevalier et al. (2004), turbines in a nuclear plant.

Reviewing the above-referenced CBM optimization studies that address the smart interpretation of CM signals, it is clear that more CM data than are really necessary are usually acquired by an organization. In these studies it has often been possible to obtain a good understanding of the most important CM measurements associated with identifying the risk of equipment failing. This is achieved through a careful analysis of the data acquired by the CM specialists, along with information contained in work orders.

A side benefit of homing in on the key measurement as a result of the optimization approach is that it may be possible to reduce the number of measurements taken at the time of condition monitoring. However, care needs to be taken if measurements are discontinued since the PHM is applicable to the operating environment from which the data were acquired. If the operating environment changes, for example, due to a change in maintenance or operating practices, then perhaps the identified risk factors will no longer hold true.

Nevertheless, in a communication, Kingsbury (1999) stated in the context of discussions with United Space Alliance, the maintenance contractors for the U.S. Shuttle program:

> ... should emphasize the ability [of the CBM approach presented] to allow them to select the signals they monitor and eliminate unnecessary transducers and signal transmission or telemetry requirements. That translates into reduced weight in the orbiter and less signal band width taken up with equipment health monitoring telemetry.

A common concern raised about the use of formal statistical methods in CBM is the view that to estimate the failure distribution of an item, time-to-failure data are required. The point is that if CM is effective, then no failure will be observed, and so formal statistical procedures are impracticable. It is clear that the goal of CM is to spot when an item is about to fail, and then be proactive and take preventive action, thus preventing the failure. However, careful analysis of several sets of data has demonstrated that while the item is removed before failure, many times the removal is premature and much useful life of the item is wasted.

To elaborate, in a study of Pratt & Whitney (P&W) engines on the Boeing 707 (Anderson et al., 1982), while most engines (42 of 50) in the sample survived their design life, of the 50 examined, 8 had been removed prior to the end of their design life due to readings from oil analysis and sent to P&W for engine overhaul. Of the eight, the maintenance reports indicated that four were good removals, but the other four were premature removals (i.e., only 50% of the removals were good removals). In fact, this reality is what prompted the early work on the possible use of the statistical procedure of PHM as an attempt to get a good handle on the real risk of an item subject to CM failing.

In a study reported by Wiseman (2001) on the optimization of CBM decisions relating to wheel motors in a mining company, no catastrophic failure of wheel motors was recorded.

Clearly, the purpose of CM is to mitigate the consequences of failures. However, in the context of optimization, one is always examining tradeoffs. So, while the outcome of CM may result in a substantial reduction in the number of failures that may have been experienced prior to the implementation of CM, perhaps down to zero, a question that could be asked is: Is this reduction economically justifiable?

Unquestionably, CM does substantially improve plant reliability, but it is observed that there are often significant premature removals due to misinterpretation of the signals that emanate from various forms of CM. In the wheel motor study (Wiseman, 2001) there were many CM records associated with the 138 wheel motors in a fleet of haul trucks in an open pit mine. Oil analysis was used to monitor the health of the wheel motors, and rules were used to decide when the wheel motor should be removed. No wheel motor was removed due to unexpected failure while in operation; 94 wheel motors were removed due to CM readings. On examining the maintenance reports associated with the rebuilds, it was identified that 32 of the motors could be classed as failures, that is, had been removed shortly before one might have expected a failure. The other 62 could be classed as premature removals; they had useful life left in them and could have safely been left in service. So when building the PHM for wheel motors, the 32 good removals were treated as failures and the other 62 as suspensions.

A final comment: In the RCM literature, much comment is made about the fact that when a study was undertaken on civil aircraft (Moubray, 1991), most failures of equipment could be described by a hazard (risk of failure, sometimes called conditional probability of failure) that was constant. At the time of the study referred to by Moubray, only the time to failure was measured. For complex items it is to be expected that the hazard will be constant, since failures can arise from many different causes, thus appearing completely random, following a Poisson process. For example, in the petrochemical industry, where simulation was used to establish maintenance crew sizes and shift patterns, β of the fitted Weibull distribution was 1 for items, including cement makeup system, halogenator, coagulator, baler, conveyors, wrapper, crusher, and so on, and therefore hazard is constant (Saint-Martin, 1985). In a 25-year-old thermal generating station where simulation was used to establish how best to improved plant performance through refurbishment, β of the fitted Weibull distribution was found to be 1 for pulverizers, gas system, waterwalls, economizer, turbines, transformer, circulating pumps, and so forth (Concannon et al., 1990).

It should be noted that in the above two examples, the risk of failure was only estimated on the basis of one key measurement — working age. In the case of the petrochemical plant, it was tons. For the power station, it was operating hours.

However, as can be observed from this section, if other measurements in addition to age are being obtained and used in the hazard rate calculation, using the PHM, then the age of an item may well be identified as having an influence on its hazard rate.

Referring back to Chapter 2, if the item being examined is a line replaceable unit (LRU) — one where the only maintenance action taken is equivalent to renewal of the item, either preventively or on failure — and the hazard rate is constant, then age has no influence on the hazard function and the optimal replacement time is infinity, that is, replace only on failure (R-o-o-F).

3.5.6 Software for CBM Optimization

To take advantage of the theory described in Section 3.5.3, a software package named EXAKT (www.omdec.com) has been developed. As explained by Wiseman (2004), "EXAKT takes processed signals, correlates them with past failure and potential failure events. Using modeling, it subsequently provides failure risk and

TABLE 3.5
Vibration-Monitoring Data

Ident	Date	WorkingAge	OVER_A	VEL_1A	VEL_2A	VEL_3A	VEL_4A
N2	7/21/95	1	0.383	0.37	0.025	0.011	0.046
N2	7/22/95	2	0.422	0.406	0.022	0.054	0.09
N2	7/23/95	3	0.067	0.057	0.023	0.013	0.02
N2	8/21/95	32	0.282	0.275	0.024	0.034	0.018
N2	9/20/95	62	0.15	0.14	0.027	0.0088	0.037
N2	9/28/95	70	0.069	0.059	0.027	0.013	0.018
N2	10/25/95	97	0.074	0.067	0.028	0.011	0.0091
N2	11/22/95	125	0.037	0.027	0.022	0.0098	0.0045
N2	12/20/95	153	0.275	0.269	0.02	0.018	0.043
N3	1/24/96	35	0.224	0.218	0.025	0.0073	0.038
N3	2/12/96	53	0.067	0.056	0.029	0.019	0.0049
N3	2/29/96	70	0.104	0.02	0.014	0.019	0.08
N3	3/04/96	74	0.305	0.161	0.255	0.032	0.0029
N3	5/22/96	152	0.052	0.023	0.041	0.01	0.019
N3	5/31/96	161	0.089	0.085	0.016	0.0052	0.022
N3	6/12/96	173	0.059	0.033	0.013	0.009	0.043
N3	6/24/96	185	0.043	0.019	0.0078	0.0045	0.036
N3	7/17/96	208	0.049	0.0094	0.013	0.034	0.026
N3	8/16/96	238	0.185	0.0091	0.0075	0.0026	0.178
N4	8/26/96	10	0.125	0.015	0.124	0.0091	0.0019
N4	9/20/96	35	0.109	0.008	0.011	0.0031	0.0036
N4	1/30/97	167	0.07	0.0076	0.0034	0.0029	0.0032
N4	5/14/97	271	0.064	0.011	0.026	0.0027	0.0074
N4	5/22/97	279	0.057	0.011	0.0036	0.004	0.0078
N4	6/19/97	307	0.164	0.011	0.0066	0.012	0.045
N4	7/15/97	333	0.104	0.0077	0.0075	0.0067	0.023
N4	9/01/97	380	0.212	0.012	0.023	0.0088	0.047

residual life estimates tuned to the economic considerations and the availability requirements for that asset in its current operating context."

Table 3.5 illustrates CM data that EXAKT requires if the CM tool is vibration monitoring, such as equipment identification number; age of item at inspection; vibration measurements, such as overall acceleration; velocity in the vertical direction, first bandwidth; velocity in the vertical direction, second bandwidth, and so forth. In addition, event data are required. This is information about when equipment went into service and when it came out of service, and whether removal was preventive or on failure. It is also information about any maintenance interventions that took place between installation and removal of the equipment, which may affect interpretation of the CM data, such as the events defined in Table 3.6. A sample of the vibration analysis event data for the example illustrated in this section is provided in Table 3.7, where the working age is days.

3.5.6.1 Definition of an Event

1. A beginning event. This indicates the start of a history. (A history includes all events from installation to removal of an item.) Designated by B.
2. A failure event. Designated by EF (ending with failure).
3. A preventive replacement. Designated by ES (ending by suspension).

TABLE 3.6
Different Forms of Event Data

1. An oil change
2. A rotor balance
3. A shaft/coupling alignment
4. A soft foot correction
5. Tightening, calibration, minor adjustments that affect the condition data
6. A filter replacement

TABLE 3.7
Vibration Analysis Event Data

Ident	Date	WorkingAge	Event
B1	9/28/94	0	B
B1	11/26/94	59	EF
B2	11/26/94	0	B
B2	1/12/95	47	EF
B3	1/12/95	0	B
B3	7/20/95	189	EF
B4	7/20/95	0	B
B4	8/21/95	33	EF
B5	8/21/95	0	B
B5	12/20/95	121	EF
B6	12/20/95	0	B
B6	3/04/96	75	EF
B7	3/04/96	0	B
B7	7/15/97	497	EF

1 / 55

An event is also an occurrence during a history that affects the condition data. Some examples are Table 3.6 and Table 3.7.

Data from Table 3.5 and Table 3.7 (only parts of the complete tables are shown here) are used to obtain the PHM. The same data are used to estimate the probability of going from one state of the vibration measurement to another state during a specified interval, known as a transition probability, which is then used in combination with cost data to obtain the optimal decision figure (see Banjevic et al., 2001). Table 3.8 is an example of the transition probability matrix for the vibration measurement "velocity in the axial direction, first bandwidth" and when the interval for the transition is specified as 30 days. Thus, if today the velocity is in the range of 0.15 to 0.22, there is a probability of 0.37788 that the equipment will be in the same state 30 days from today. Similarly, the table can be used to find the probability of the equipment being in a failure state in 30 days' time as 0.199714. Transition probabilities are provided for all possible combinations of states.

TABLE 3.8
Transition Probability Matrix

Inspection Interval = 30 days

	VEL#1A Age 0 to 180 (days)	0 to 0.1	0.1 to 0.15	0.15 to 0.22	0.22 to 0.37	Above 0.37
Very Smooth	0 to 0.1	0.5754	0.2242	0.1452	0.0405	0.0147
Smooth	0.1 to 0.15	0.2059	0.2498	0.3309	0.1374	0.0760
Rough	0.15 to 0.22	0.0554	0.1376	0.3779	0.2294	0.1997
Very Rough	0.22 to 0.37	0.0129	0.0474	0.1904	0.2424	0.5069
Failure	Above 0.37	0.0005	0.0027	0.0170	0.0521	0.9277

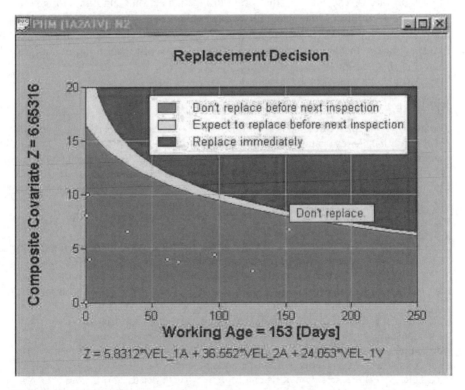

FIGURE 3.15 Optimizing the CBM decision.

Finally, using the PHM, transition probabilities, and the costs associated with preventive and failure replacement, the graph that can be used for decision making is obtained (Figure 3.15).

Thus, whenever an inspection is made, the values of the key risk factors are obtained. In this case, the key risk factors are velocity in the axial direction, first bandwidth; velocity in the axial direction, second bandwidth; and velocity in the vertical direction, first bandwidth. These measurements are then multiplied by their weighting factors, 5.8312, 36.552, and 24.053, respectively, and then added together to give a Z value, which is marked on the Y-axis. The X-axis shows the age of the item (a bearing in this example) at the time of inspection. The position of the point on the graph indicates the optimal decision. If the point is in the lightly shaded area, the recommendation is to continue operating — with reference to the risk plot in Figure 3.15, the hazard is below the optimal level. If the intersection is in the dark-shaded area, the recommendation is to replace — in this case, the hazard is greater than the optimal risk level. If the intersection lies in the clear area, it indicates that the optimal change-out time is between two inspections.

On the Web site www.omdec.com there is a detailed explanation of EXAKT, along with the answers to many frequently asked questions and a number of tutorial problems. The chapter "Interpretation of Inspection Data Emanating from Equipment Condition Monitoring Tools: Method and Software" in *Mathematical and Statistical Methods in Reliability* (Jardine and Banjevic, 2005) provides an overview of the theory and application of the CBM optimization approach presented in this section.

REFERENCES

Anderson, M., Jardine, A.K.S., and Higgins, R.T. (1982). The use of concomitant variables in reliability estimation. *Modeling and Simulation*, 13, 73–81.

Banjevic, D., Jardine, A.K.S., Makis, V., and Ennis, M. (2001). A control-limit policy and software for condition-based maintenance optimization. *INFOR*, 39, 32–50.

Chevalier, R., Benas, J.-C., Garnero, M.A., Montgomery, N., Banjevic, D., and Jardine, A.K.S. (2004). Optimizing CM Data from EDF Main Rotating Equipment Using Proportional Hazard Model. Paper presented at Surveillance5 Conference, Senlis, France, October 11–13, 2004.

Christer, A.H. (1973). Innovatory decision making. In *The Role and Effectiveness of Theories of Decision in Practice*, D.J. White and K.C. Bowen, Eds. Hodder and Stoughton.

Christer, A.H. (1999). Developments in delay time analysis for modelling plant maintenance. *Journal of the Operational Research Society*, 1120–1137.

Christer, A.H. and Waller, W.M. (1984). Delay time models of industrial inspection maintenance problems. *Journal of the Operational Research Society*, 35, 401–406.

Concannon, K.H., McPhee, J., and Jardine, A.K.S. (1990). Balance maintenance costs against plant reliability with simulation modelling. *Industrial Enginering*, 24–27.

Cox, D.R. (1972). Regression models and life tables (with discussion). *J. Roy. Stat. Soc. B*, 34, 187–220.

Drenick, R.F. (1960). The failure law of complex equipment. *Journal of the Society of Industrial and Applied Mathematics*, 8, 680–690.

Jardine, A.K.S. (2002). Optimizing condition based maintenance decisions. In *Proceedings, Reliability and Maintainability Symposium*, pp. 90–97.

Jardine, A.K.S. and Banjevic, D. (2005). Interpretation of inspection data emanating from equipment condition monitoring tools: method and software. In *Mathematical and Statistical Methods in Reliability*, Y.M. Armijo, Ed. World Scientific Publishing Company, Singapore.

Jardine, A.K.S., Banjevic, D., Wiseman, M., and Buck, S. (2001). Optimizing a mine haul truck wheel motors' condition monitoring program. *Journal of Quality in Maintenance Engineering*, 1, 286–301.

Jardine, A.K.S. and Hassounah, M.I. (1990). An optimal vehicle-fleet inspection schedule. *Journal of the Operational Research Society*, 41, 791–799.

Jardine, A.K.S., Joseph, T., and Banjevic, D. (1999). Optimizing condition-based maintenance decisions for equipment subject to vibration monitoring. *Journal of Quality in Maintenance Engineering*, 5, 192–202.

Jardine, A.K.S., Kahn, K., Banjevic, D., Wiseman, M., and Lin, D. (2003). An Optimized Policy for the Interpretation of Inspection Data from a CBM Program at a Nuclear Reactor Station. Paper presented at COMADEM, Sweden, August 27–29.

Jeffris, T., Banjevic, D., Jardine, A.K.S., and Montgomery, N. (2004). Oil Analysis of Marine Diesel Engines: Optimizing Condition-Based Maintenance Decisions. Paper presented at COMADEM 2004, Robinson College, Cambridge, England, August 23–25.

Jia, X. and Christer, A.H.C. (2003). Case experience comparing the RCM approach to plane maintenance with a modeling approach. In *Case Studies in Reliability and Maintenance*, W.R. Blische and D.N. Murthy, Eds. New York: Wiley.

Jiang, R. and Jardine, A.K.S. (2005). Two optimization models of the optimum inspection problem. *Journal of the Operational Research Society*.

Kalbfleisch, J.D. and Prentice, R.L. (2002). *The Statistical Analysis of Failure Time Data*, 2nd ed. New York: Wiley.

Kingsbury, R.L. (1999). Private communication subsequent to presentation: Kingsbury, R.L., Jardine, A.K.S., and Picknell, J.V. (1998). Reliability Centered Maintenance. Paper presented at Aerospace Technology Working Group, NASA/Industry National Meeting, Cocoa Beach, FL, October 14–16.

Lin, D., Lally, R., Wiseman, M., Banjevic, D., and Jardine, A.K.S. (2003). An Automated Prognostic Intelligent Agent for the AAAV Using Proportional Hazard Modeling. Paper presented at Proceedings, The JANNAF 39th CS, 27th APS, 21st PSHS and 3rd MSS Joint Meeting, Colorado Springs, CO, December 1–5.

Makis, V. and Jardine, A.K.S. (1992). Optimal replacement in the proportional hazards model. *INFOR*, 20, 172–183.

Monnot, M., Heston, S., Banjevic, D., Montgomery, N., and Jardine, A.K.S. (2004). Smartly Interpreting Oil Analysis Results from the Hydraulic System of Backhoes. Paper presented at STLE Conference, Toronto, May 17–19.

Moubray, J.M. (1991). *Reliability-Centered Maintenance* II. Oxford: Butterworth-Heinemann.

Moubray, J.M. (1997). *Reliability-Centered Maintenance* II, 2nd ed. Oxford: Butterworth-Heinemann, p. 172.

O'Connor, P.D.T. (2002). *Practical Reliability Engineering*, 4th ed. New York: Wiley.

O'Hanlon, T. (2003). *Maintenance Technology*, December, p. 54.

Saint-Martin, J.R.C. (1985). Computer Simulation of a Chemical Processing Plant. M.Eng thesis, Royal Military College of Canada, Kingston, Ontario.

Vlok, P.J., Coetzee, J.L., Banjevic, D., Jardine, A.K.S., and Makis, V. (2002). Optimal component replacement decisions using vibration monitoring and the PHM. *Journal of the Operational Research Society*, 53, 193–202.

White, D.J., Donaldson, W.A., and Lawrie, N.L. (1969). *Operational Research Techniques*, Vol. 1. London: Business Books.

Willets, R., Starr, A.G., Banjevic, D., Jardine, A.K.S., and Doyle, A. (2001). Optimizing Complex CBM Decisions Using Hybrid Fusion Methods. Paper presented at COMADEM, University of Manchester, Manchester, U.K., September.

Wiseman, M. (2001). Optimizing condition based maintenance. In *Maintenance Excellence*, J.D. Campbell and A.K.S. Jardine, Eds. New York: Marcel Dekker, chap. 12.

Wiseman, M. (2004). Private communication.

PROBLEMS

1. The current maintenance policy being adopted for a complex transfer machine in continuous operation is that inspections are made once every 4 weeks. Any potential defects that are detected during this inspection and that may cause breakdown of the machine are rectified at the same time. In between these inspections, the machine can break down, and if it does so, it is repaired immediately. As a result of the current inspection policy, the mean time between breakdowns is 8 weeks.

   It is known that the breakdown rate of the machine can be influenced by the weekly inspection frequency, n, and associated minor maintenance undertaken after the inspection, and is of the form $\lambda(n) = K/n$, where $\lambda(n)$ is the mean rate of breakdowns per week for an inspection frequency of n per week.

   Each breakdown takes an average of 1/4 week to rectify, while the time required to inspect and make minor changes is 1/8 week.

   a. Construct a mathematical model that could be used to determine the optimal inspection frequency to maximize the availability of the transfer machine.

   b. Using the model constructed in (a) along with the data given in the problem statement, determine the optimal inspection frequency. Also give the availability associated with this frequency.

2. An industrial machine consists of two parts, part A and part B. Each part has its own rate of breakdown, and whenever either of the two parts breaks down, the whole machine will be stopped for repair. Each breakdown takes an average of 3 days to rectify. At inspections, both parts A and B are inspected, and in total, the inspection takes 1.5 days to complete. Any potential defects that are detected during these inspections and that may cause breakdown of the machine are rectified at the time. It is known that the breakdown rates of parts A and B are influenced by the inspection frequency, n, and associated minor maintenance work, and they are of the form listed below:

$$\text{Part A: } \lambda_1(n) = \frac{K_1}{n}$$

$$\text{Part B: } \lambda_2(n) = \frac{K_2}{n}$$

where $\lambda_1(n)$ and $\lambda_2(n)$ are the mean rates of breakdowns per month for parts A and B, respectively, when an inspection frequency of n per month applies.

The current maintenance policy being adopted for the machine in continuous operation is that inspections are made once a month. As a result of this policy, the mean time between machine breakdowns is 2 months.

a. Construct a mathematical model that could be used to determine the optimal inspection frequency to maximize the availability of the machine.
b. Using the model constructed in (a) along with the data given in the problem statement, determine the optimal inspection frequency. Also give the availability associated with this frequency.
c. Find the value of K_1 given that $K_2 = 0.1$. (The calculations have to consider a month as a unit of time.)

3. Consider the pumps shown in Figure 3.16. The duty pump (pump B) is pumping water into a tank (tank Y) from which the water is drawn at a rate of 800 l/min to cool a reactor that is working continuously 24 hours a day, 7 days a week. The duty pump is switched on by one float switch when the level in tank Y drops to 120,000 l, and switched off by another when the level reaches 240,000 l. A third switch is located just below the low-level switch of the duty pump. This switch is

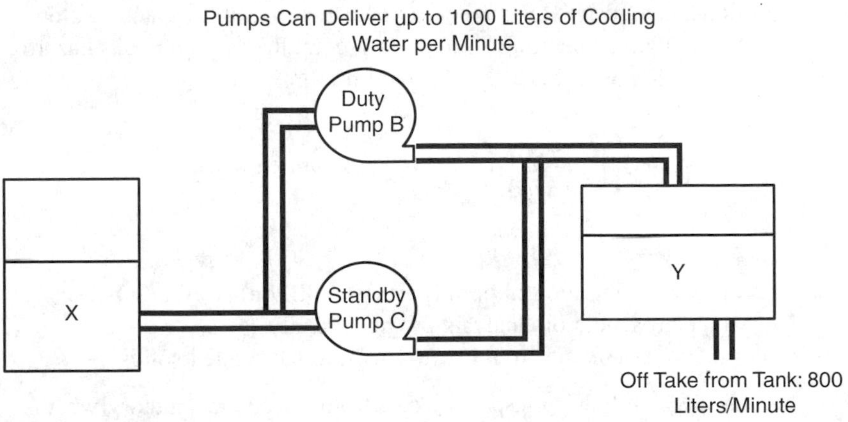

FIGURE 3.16 A cooling water supply system.

designed both to sound an alarm in the control room if the water level reaches it, and to switch on the standby pump (pump C). If the tank runs dry, which happens when the standby pump is in a failed state while it is required to pump the cooling water, the reactor has to be shut down. If it is not shut down, there will be no cooling water for the reactor, which means there is a high probability of a catastrophic failure. The probability of having an explosion when no cooling water is circulating in the reactor and before the operators shut down the reactor is 10%. According to governmental regulations, the plant has to ensure that the probability of having such an explosion in each year does not exceed 10^{-8}. The mean time between failures for the duty pump is 24 months. The standby pump can fail while it is idle with the rate of one failure every 120 months. Currently the operators turn on the standby pump every few months to check whether it is still capable of working. The maintenance department has done some calculations based on the above information and concluded that in order not to breach governmental regulations, the standby pump has to have an availability of $>(1 - 2 \times 10^{-5})$.

a. Determine the mean time between each inspection of the standby pump in order to provide the availability suggested by the maintenance department.

b. Is your answer to part (a) feasible? Why? If the answer is no, make some suggestions on how you can reach such availability.

c. In fact, the calculation done by the maintenance department is wrong. Calculate the correct required availability for the standby pump (Figure 3.16).

4. Rather than basing component hazard rate predictions solely on accumulated utilization, it may be possible to use concomitant information to improve predictions. The following model, derived from Cox's proportional hazards model, includes explanatory variables z_1 and z_2, along with cumulative operating hours, t, to predict the instantaneous hazard rate, $h(t)$, for wheel motors of a haul truck.

$$h(t, z_1, z_2) = \frac{2.891}{23,360} \left(\frac{t}{23,360} \right)^{1.891} e^{(0.002742 z_1 + 0.0000539 z_2)}$$

where
z_1 = iron concentration in ppm (parts per million) from a SOAP (spectroscopic oil analysis program) analysis
z_2 = sediment reading from test to measure suspended solids

a. Given the following inspection data from three wheel motors, what is your estimate of the current hazard rate for each motor (Table 3.9)?

TABLE 3.9
Inspection Data from Three Wheel Motors

Wheel Motor Number	Age (hours)	Iron (ppm)	Sediment Measurement
1	11,770	5	6.0
2	11,660	2	6.0
3	8460	12	2.4

  b. You are asked to submit a report to mine maintenance regarding the hazard value for motor 1. How might you explain the value you have obtained, and what maintenance action would you recommend given that a maintenance stoppage is scheduled in 10 days?
  c. How would you interpret the number 2.891 in the hazard model?

5. A company monitors the gearboxes on vehicles by attaching a wireless sensor to each gearbox to take vibration readings. The vibration signals are then analyzed by a digital signal processing toolbox. Two condition indicators showing the health of the gearbox, CI1 and CI2, are extracted from each vibration signal. After running the above condition monitoring on a fleet of vehicles, the company has accumulated a certain amount of data. Now the company manager asks you to apply EXAKT to the data.

  a. The first step for you would be to collect and prepare the data. What are the two main sources of data required by EXAKT?
  b. Now you have obtained the right data and have properly prepared it. You want to establish a PHM for the gearboxes. The usual way for the modeling is to include both indicators in the PHM.
    i. If you find one of them, CI1, is significant and the other, CI2, is not, how would you proceed with the modeling? What would you do if both CI1 and CI2 are not significant?
    ii. If you find that the shape parameter is not significant (i.e., $\beta = 1$), how would you proceed? What does it really mean when you say the shape parameter is *not significant*?
  c. Assume that the final PHM you get is

$$h(t, \text{CI2}) = \left(\frac{5.4184}{10319}\right)\left(\frac{t}{10319}\right)^{4.4184} \exp\left(0.3884\text{CI2}\right)$$

  where $h(t, \text{CI2})$ is the hazard rate and t is the operation hours.
    Given the following data from three gearboxes, estimate the hazard rate of each gearbox (Table 3.10).

TABLE 3.10
Data from Three Gearboxes

Gearbox No.	Operation Hours	CI1	CI2
1	8550	2	5.5
2	3215	10	2.1
3	9460	12	1.4

d. You are asked to submit a report regarding the hazard rate of gearbox 1. How might you explain the value you have obtained and what maintenance action would you recommend within the next 48 hours?

4 Capital Equipment Replacement Decisions

First weigh the considerations, then take the risks.

Helmuth von Moltke

4.1 INTRODUCTION

The goal of this chapter is to present models that can be used to determine optimal replacement decisions associated with capital equipment by addressing life cycle costing (LCC) decisions, or its complement, life cycle profit (LCP), sometimes termed whole-life costing (WLC). Capital equipment problems tend to be treated deterministically, and that is the approach taken in this chapter.

In the context of the framework of the decision areas addressed in this book, we are addressing column 3 of the framework, as highlighted in Figure 4.1.

Four classes of problem will be addressed in this chapter:

1. Establishing the economic life of equipment that is essentially utilized steadily each year
2. Establishing the economic life of equipment that has a planned varying utilization, such as using new equipment for base load operations and using older equipment to meet peak demands
3. Deciding whether to replace present equipment with technologically superior equipment, and if so, when
4. Deciding on the best action: repair (rebuild) vs. replace.

The basic issue to be addressed in each case is illustrated in Figure 4.2.

From Figure 4.2 we can see that as the replacement age of an item increases, the operations and maintenance costs per unit time will increase, while the ownership cost will decrease. In simple terms, the ownership cost is the purchase price of an asset minus its resale value at the time of replacement, divided by the replacement age. There may be additional costs incurred associated with the utilization of an item that are independent of the age at which the asset is replaced. These are identified as the fixed costs. The economic life (optimum replacement age) is then that time at which the total cost, in terms of its equivalent annual cost (EAC) — see Appendix 3 — is at its minimum value.

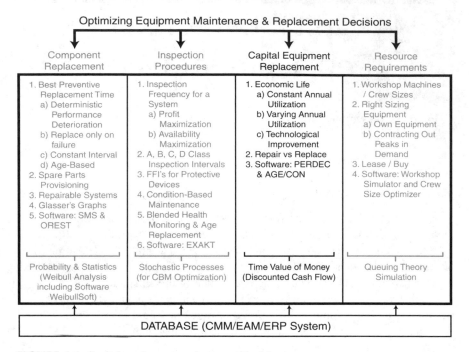

FIGURE 4.1 Capital equipment replacement decisions.

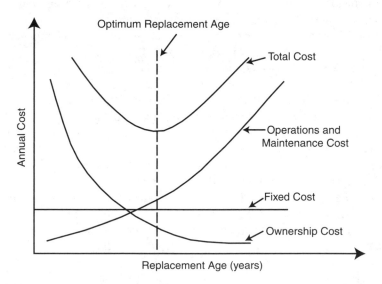

FIGURE 4.2 Classic economic life conflicts.

It will be noted from examination of Figure 4.2 that the fixed costs will not affect the economic life decision, so they can be omitted from the analysis. However, when finalizing budget requirements for replacing assets at the end of their economic life, it is necessary to remember to include the fixed costs. They are part of the budget.

4.2 OPTIMAL REPLACEMENT INTERVAL FOR CAPITAL EQUIPMENT: MINIMIZATION OF TOTAL COST

4.2.1 STATEMENT OF PROBLEM

Through use, equipment deteriorates and this deterioration may be measured by an increase in the operations and maintenance (O&M) costs. Eventually the O&M costs will reach a stage where it becomes economically justifiable to replace the equipment. What we wish to determine is an optimal replacement policy that minimizes the total discounted costs derived from operating, maintaining, and disposing of the equipment over a long period. It will be assumed that equipment is replaced by identical equipment, thus returning the equipment to the as-new condition after replacement. (This restriction will be relaxed in the problem of Section 4.5 when dealing with technological improvement.) Furthermore, it is assumed that the trends in O&M costs following each replacement will remain identical. Since the equipment is being operated over a long period, the replacement policy will be periodic, and so we will determine the optimal replacement interval.

4.2.2 CONSTRUCTION OF MODEL

1. A is the acquisition cost of the capital equipment.
2. C_i is the operation and maintenance cost in the ith period from new, assumed to be paid at the end of the period, $i = 1, 2, ..., n$.
3. S_i is the resale value of the equipment at the end of the ith period of operation, $i = 1, 2, ..., n$.
4. r is the discount factor (for details, see Appendix 3).
5. n is the age in periods (such as years) of the equipment when replaced.
6. $C(n)$ is the total discounted cost of operating, maintaining, and replacing the equipment (with identical equipment) over a long period, with replacements occurring at intervals of n periods.
7. The objective is to determine the optimal interval between replacements to minimize total discounted costs, $C(n)$.

The replacement policy is illustrated in Figure 4.3.

Consider the first cycle of operation: the total discounted cost up to the end of the first cycle of operation, with equipment already purchased and installed, is

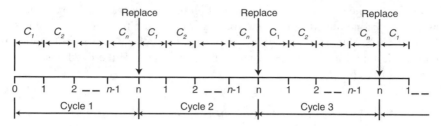

FIGURE 4.3 Optimal replacement interval: capital equipment.

$$C_1(n) = C_1 r^1 + C_2 r^2 + C_3 r^3 + \ldots + C_n r^n + A r^n - S_n r^n = \sum_{i=1}^{n} C_i r^i + r^n (A - S_n)$$

For the second cycle, the total cost discounted to the start of the second cycle is

$$C_2(n) = \sum_{i=1}^{n} C_i r^i + r^n (A - S_n)$$

Similarly, the total costs of the third cycle, fourth cycle, and so forth, discounted back to the start of their respective cycle, can be obtained.

The total discounted costs, when discounting is calculated at the start of the operation, time 0, is

$$C(n) = C_1(n) + C_2(n) r^n + C_3(n) r^{2n} + \ldots + C_n(n) r^{(n-1)n} + \ldots$$

Since $C_1(n) = C_2(n) = C_3(n) = \ldots = C_n(n) = \ldots$, we have a geometric progression that gives, over an infinite period,

$$C(n) = \frac{C_1(n)}{1 - r^n} = \frac{\displaystyle\sum_{i=1}^{n} C_i r^i + r^n (A - S_n)}{1 - r^n} \tag{4.1}$$

This model of the problem relates replacement interval n to total costs.

4.2.3 NUMERICAL EXAMPLE

1. Let $A = \$5000$.
2. The estimated operations and maintenance costs per year for the next 5 years are in Table 4.1.
3. The estimated resale values over the next 5 years are in Table 4.2.
4. The discount factor $r = 0.9$.

TABLE 4.1
Trend in O&M Costs

Year	1	2	3	4	5
Estimated O&M Cost ($)	500	1000	2000	3000	4000

TABLE 4.2
Trend in Resale Values

Year	1	2	3	4	5
Resale Value ($)	3000	2000	1000	750	500

TABLE 4.3
Total Discounted Costs

Replacement Time, n	1	2	3	4	5
Total Discounted Cost, $C(n)$ ($)	22,500	19,421	20,790	21,735	23,701

Evaluation of Equation 4.1 for different values of n gives the figures of Table 4.3, from which it is seen that the best time to replace (in terms of the economic life of the equipment) is after the equipment has been used for 2 years with a total discounted cost of $19,421. The best policy will then be to replace that asset at intervals of 2 years "forever." Rather than present the cost associated with an infinite chain of replacement, it is more meaningful to provide the equivalent annual cost (EAC).

Recall from Appendix 3 that EAC = PV × CRF, where PV is the present value and CRF is the capital recovery factor. Thus, we get

$$\text{EAC} = \$19,421 \times 0.11 = \$2136$$

Note that CRF = i since n in the CRF equation equals infinity due to the model that has been used.

A graphical representation is provided in Figure 4.4.

4.2.4 FURTHER COMMENTS

In the model developed above (Equation 4.1) it was assumed that we had started with equipment in place and asked the question: When should it be replaced? Thus, the first time that the acquisition cost of the item is incurred is at the end of the first cycle of operation. Furthermore, it was assumed that the O&M costs were incurred at the end of the year, and so, for example, first-year costs were

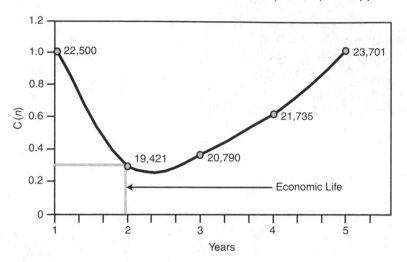

FIGURE 4.4 Total discounted cost vs. equipment age.

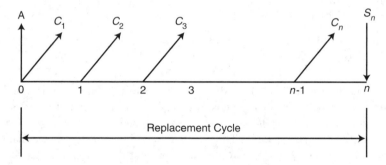

FIGURE 4.5 Purchase price at start of cycle.

discounted by 1 year. Perhaps a more realistic assumption is that the purchase price of the asset is incurred at the start of the replacement cycle, and also that costs in a year are incurred at the start of the year; therefore, for example, year 1 costs are not discounted. This is illustrated in Figure 4.5. Of course, one could assume that costs are incurred continuously during the year, and therefore continuous discounting would be used (see Appendix 3). In practice, what is usually assumed in capital equipment replacement studies is what is depicted in Figure 4.5.

Thus, in all replacement studies, one needs to be clear when the cash flows occur. The model represented by Equation 4.2 reflects the cash flows depicted in Figure 4.5 and is what is used in the economic life software PERDEC and AGE/CON, presented in Section 4.7. Again, it is the EAC that is calculated:

$$\text{EAC}(n) = \frac{A + \sum_{i=1}^{n} C_i r^{i-1} - r^n S_n}{1 - r^n} \times i \tag{4.2}$$

In the models discussed it has been assumed that the acquisition cost of equipment remains constant. Also, it is assumed that the trend in maintenance costs is the same after each replacement. Because of inflationary trends, this is unlikely, and therefore, it may be necessary to modify the model to take account of these facts. However, as explained in Appendix 3, provided an appropriate interest rate is used, inflation effects do not need to be incorporated into the model. However, if required, it can be done.

In the models for capital equipment replacement no consideration was given to tax allowances that may be available. This is an aspect that is rarely mentioned in replacement studies, but which must be included where relevant. One paper that extends the models discussed here to consider issues of tax is that of Christer and Waller (1987).

It may seem unreasonable that we should sum the terms of the geometric progression to infinity. This, however, does make the calculations a little easier and ensures that all replacement cycles are compared over the same period.

Note, however, that rather than use the model, either Equation 4.1 or Equation 4.2, it is possible just to calculate the EAC for different replacement cycles, and this will result in the same economic life being identified. The model then is

$$\text{Economic life} = \text{Min} \left[A + \sum_{i=1}^{n} C_i r^{i-1} - r^n S_n \right] \times \text{CRF}$$

Note 1: Both Equations 4.1 and 4.2 give the same result for the economic life of an asset. However, the EAC will be lower if Equation 4.1 is used.

Note 2: It can be argued that an appropriate model to use is one where the acquisition cost, A, is first introduced at the end of the replacement cycle, since if an asset is required, it must be purchased, and so the purchase price can then be considered a "sunk" cost, and so the decision to be made is to establish the economic time to replace that item with a new one, taking into account the accumulated operations and maintenance costs and the purchase price of a new asset.

4.2.5 Applications

4.2.5.1 Mobile Equipment: Vehicle Fleet Replacement

The policy in place in a trucking fleet was to replace the vehicles on a 5-year cycle. The question was asked: What is the economic life of the vehicles used in a fleet of 17 units?

The data available included a purchase price of $85,000. Trends in O&M costs, resale values, and interest rate for discounting are given in Table 4.4.

Figure 4.6 provides the results, from which it is seen that the economic life is 1 year, with an EAC of about $65,000. (Note that the model used to obtain the EACs on Table 4.4 was Equation 4.2.)

TABLE 4.4
Vehicle Fleet Data

Age of Vehicle (years)	O&M Cost (in today's $)	Rate for Cash Flow Discounting (5)	Trade-In Value ($)
1	29,352	10	60,000
2	45,246	10	40,000
3	52,626	10	25,000
4	53,324	10	20,000
5	42,363	10	15,000

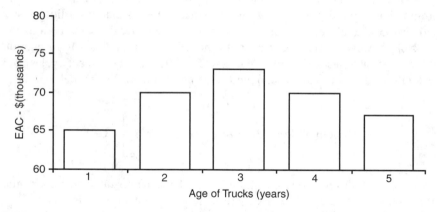

FIGURE 4.6 Vehicle fleet economic life.

That the economic life was identified as 1 year came as quite a surprise to the organization. However, when many maintenance problems are addressed from a data-driven solution perspective, surprises do occur. Intuition and common/past practices do not always provide good solutions.

Examination of Figure 4.6 raises some interesting questions, such as: By increasing the age at which the vehicle is replaced, would the EAC be lower in year 6 or year 7? Why is the economic life 1 year? To answer such questions, it is always appropriate to "look behind the statistics." In this case, there were two reasons why the economic life was identified as 1 year:

1. The substantial increase in O&M costs in year 2 compared with year 1
2. The high resale value associated with a 1-year-old vehicle

Before proceeding to implement a 1-year cycle, it is necessary to know if there are any concerns about these values. Asking why a few times usually gets to the root of the underlying cause. In this example, the cause of a 1-year answer was due to a large number of warranty claims being accepted in year 1. Also, an estimate of $60,000 was provided by the vehicle fleet supplier as a trade-in for a 1-year-old vehicle, but will this be the case if the fleet operator changes from the current practice of replacing the vehicles on a 5-year cycle to a 1-year cycle? If this is done, then the supplier will be receiving 17 vehicles every year. Once the supplier realizes the implication, then there may be a reduction in the estimate of the value of a 1-year-old vehicle. If this happens, perhaps the economic life will change. Thus, there is good reason for the fleet operator to undertake a sensitivity analysis on the effect of a decreased resale value on the economic life of the truck.

Figure 4.6 might suggest that a replacement at year 6 would give a cost less than that at year 5, and perhaps less than that at year 1. Why is the EAC decreasing? Is it because of a major maintenance action in year 3, and benefits are being realized in years 4 and 5? This may be a possible answer. In the study, however, the reason for the decreasing EAC is that given the established practice of replacing the vehicles on a 5-year cycle, any avoidable maintenance cost in year 5 was not incurred. If a decision was made to increase the life past 5 years, then additional maintenance costs would be incurred in years 4 and 5.

4.2.5.2 Fixed Equipment: Internal Combustion Engine

The organization was planning to purchase four new combustion engines and wanted to know what their expected economic life might be. In addition, there was an alternative engine that could be purchased, so the question then became: What is the best buy?

The data for engine A included a purchase and installation cost of $19 million. O&M costs were estimated for the next 15 years by judicious use of manufacturer's data along with data contained in a database used by the oil and gas industry. Much sensitivity checking was undertaken to obtain a robust trend in O&M costs. Similarly, an estimate of the trend in resale values was obtained — and for specialized equipment that resale value may be a scrap value or could even be zero, no matter when the asset is replaced. If that is the case, then $S_i = 0$ for all replacement ages. The interest rate appropriate for discounting was provided by the company. Calculating the EAC for the 15 years for which data were available gave Figure 4.7, from which it is seen that the EAC is still declining, and no minimum has been identified. However, one can conclude that we are close to the minimum, and at 15 years the EAC is $5.36 million.

The data for engine B included a purchase and installation cost of $14.5 million. Similar to engine A, the O&M cost trend and resale value information was obtained, the same interest rate was used, and the resulting EAC trend is provided in Figure 4.8, which shows a pattern similar to that for engine A. The EAC at 15 years is $3.17 million.

The conclusion: For both engines, their economic life is greater than 15 years, the limit of available data. However, a major benefit of the economic life analysis is the identification of the fact that, based on the data used, engine B is a better buy since its EAC is $2.19 million lower than engine A. Over a 15-year period,

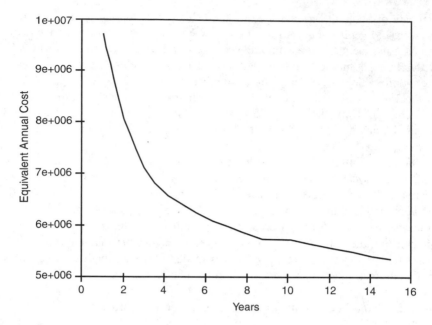

FIGURE 4.7 EAC trend for combustion engine A.

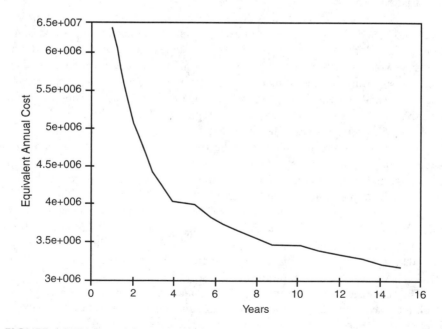

FIGURE 4.8 EAC trend for combustion engine B.

the total discounted economic benefit is $15 \times 2.19 = \$32.85$ million. The company's plan was to purchase four new combustion engines, so the economic benefit would be substantial. The solution was obtained by using a formal data-driven procedure.

4.3 OPTIMAL REPLACEMENT INTERVAL FOR CAPITAL EQUIPMENT: MAXIMIZATION OF DISCOUNTED BENEFITS

4.3.1 STATEMENT OF PROBLEM

This problem is similar to that of Section 4.2 except that (1) the objective is to determine the replacement interval that maximizes the total discounted net benefits derived from operating equipment over a long period, and (2) the trend in costs is taken to be continuous, rather than discrete.

4.3.2 CONSTRUCTION OF MODEL

1. $b(t)$ is the net benefit obtained from the equipment at time t. This will be the revenue derived from operating the equipment minus the operating costs, which may include maintenance costs, fuel costs, and so on. A possible form of $b(t)$ is illustrated in Figure 4.9.
2. $c(t)$ is the net cost of replacing equipment of age t. Replacing the equipment includes the purchase price plus installation cost, and may also include a cost for loss of production due to the time required to replace the equipment. These costs are often partially offset by the salvage value of the used equipment, which usually depends on the age of the equipment when it is replaced. A possible form of $c(t)$ is illustrated in Figure 4.10.

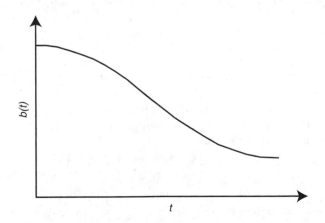

FIGURE 4.9 Net benefit trend.

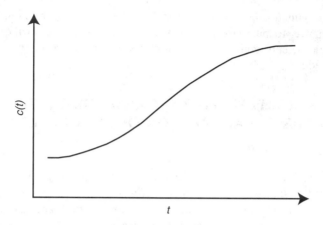

FIGURE 4.10 Trend in net cost of asset replacement.

3. T_r is the time required to replace the equipment.
4. t_r is the age of the equipment when replacement commences.
5. $t_r + T_r$ is the replacement cycle, that is, the time from the end of one replacement action to the end of the next replacement action.
6. $B(t_r)$ is the total discounted net benefits derived from operating the equipment for periods of length t_r over a long time.
7. The objective is to determine the optimum interval between replacements to maximize the total discounted net benefits derived from operating and maintaining the equipment over a long period.

$B(t_r)$ is the sum of the discounted net benefits from each replacement cycle over a long period. For purposes of the analysis, the period over which replacements will occur will be taken as infinity, although in practice this will not be the case.

4.3.2.1 First Cycle of Operation

Defining $B_1(t_r + T_r)$ as the total net benefits derived from replacing the equipment at age t_r, discounted back to their present-day value at the start of the first cycle, we get $B_1(t_r + T_r)$ = benefits received over the first cycle, that is, in interval (0, t_r), discounted to their present-day value minus the cost of replacing equipment of age t_r discounted to its present-day value.

This first cycle of operation is illustrated in Figure 4.11.

$$\text{Discounted benefits over the first cycle} = \int_0^{t_r} b(t)\exp[-it]dt$$

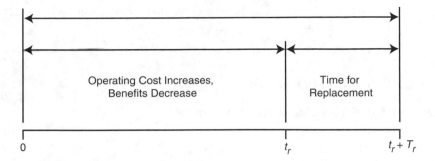

FIGURE 4.11 Replacement cycle.

where i is the relevant interest rate for discounting (see Appendix 3 for continuous discounting).

$$\text{Discounted replacement cost} = c(t_r)exp[-it_r]$$

$$B_1(t_r + T_r) = \int_0^{t_r} b(t)\exp[-it]dt - c(t_r)\exp[-it_r]$$

4.3.2.2 Second Cycle of Operation

Defining $B_2(t_r + T_r)$ as the total net benefits derived from replacing the equipment at age t_r, discounted back to their present-day value at the start of the second cycle, we get

$$B_2(t_r + T_r) = \int_0^{t_r} b(t)\exp[-it]dt - c(t_r)\exp[-it_r]$$

What we now want to do is discount $B_2(t_r + T_r)$ back to the start of the first cycle, and this is

$$B_2(t_r + T_r)\exp[-i(t_r + T_r)]$$

4.3.2.3 Third Cycle of Operation

Defining $B_3(t_r + T_r)$ as the total net benefits derived from replacing the equipment at age t_r, discounted back to give their present-day value at the start of the third cycle, we get

$$B_3(t_r + T_r) = \int_0^{t_r} b(t)\exp[-it]dt - c(t_r)\exp[-it_r]$$

Discounting $B_3(t_r + T_r)$ back to the start of the first replacement cycle, we get

$$B_3(t_r + T_r)\exp[-i2(t_r + T_r)]$$

4.3.2.4 *n*th Cycle of Operation

Defining $B_n(t_r + T_r)$ similar to the others, we get

$$B_n(t_r + T_r) = \int_0^{t_r} b(t)\exp[-it]dt - c(t_r)\exp[-it_r]$$

which discounted back to the start of the first cycle gives

$$B_n(t_r + T_r)\exp[-i(n-1)(t_r + T_r)]$$

The form that the benefits take over the first few cycles of operation is illustrated in Figure 4.12.

Thus, the total discounted net benefit, over a long period, with replacement at age t_r, is

$$B(t_r) = B_1(t_r + T_r) + B_2(t_r + T_r)\exp[-i(t_r + T_r)] + B_3(t_r + T_r)\exp[-i2(t_r + T_r)]$$

$$+ \ldots + B_n(t_r + T_r)\exp[-i(n-1)(t_r + T_r)] + \ldots$$

Since $B_1(t_r + T_r) = B_2(t_r + T_r) = B_3(t_r + T_r) = \ldots = B_n(t_r + T_r) = \ldots$, we can write

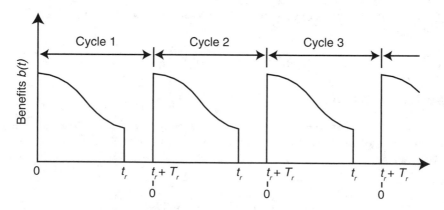

FIGURE 4.12 Discounted benefits over time.

$$B(t_r) = B_1(t_r + T_r) + B_1(t_r + T_r)\exp[-i(t_r + T_r)]$$

$$+ B_1(t_r + T_r)\exp[-i2(t_r + T_r)] + \ldots \qquad (4.3)$$

$$+ B_1(t_r + T_r)\exp[-i(n-1)(t_r + T_r)] + \ldots$$

Equation 4.3 is a geometric progression to infinity, which gives

$$B(t_r) = \frac{B_1(t_r + T_r)}{1 - \exp[-i(t_r + T_r)]}$$

That is,

$$B(t_r) = \frac{\int_0^{t_r} b(t)\exp[-it]dt - c(t_r)\exp[-it_r]}{1 - \exp[-i(t_r + T_r)]} \qquad (4.4)$$

This is a model of the replacement problem relating the replacement age t_r to the total discounted net benefits.

Rather than summing the progression to infinity, we could sum the first n terms, which gives (see Appendix 3 for formula)

$$B(t_r) = \left[\int_0^{t_r} b(t)\exp[-it]dt - c(t_r)\exp[-it_r] \right] \times \left[\frac{1 - \exp[-ni(t_r + T_r)]}{1 - \exp[-i(t_r + T_r)]} \right]$$

which results in the same optimal value of t_r as would be obtained from Equation 4.4, since the numerator in the inserted fractional expression is a constant (see proof of Section 4.3.5), although the benefit $B(t_r)$ would be reduced by this factor.

4.3.3 NUMERICAL EXAMPLE

The benefits derived from operating equipment are of the form $b(t) = \$32000e^{-0.09t}$ per year, where t is in years.

$$\text{Cost of replacement } c(t) = \$15000 - 13600e^{-0.73t}$$

The time required to perform a replacement is 1 month. Determine the optimal replacement age of the equipment when i is taken as 10% per annum.

Equation 4.4 becomes

TABLE 4.5
Economic Life: Benefit Maximization

t_r (years)	1	2	3	4	5	6	7	8
$B(t_r)$ (thousands of $)	210	232	238	239	236	232	229	225

$$B(t_r) = \frac{\int_0^{t_r} 32000e^{-0.09t}e^{-0.1t}\,dt - (15000 - 13600e^{-0.73t_r})e^{-0.1t_r}}{1 - e^{-0.1(t_r + 0.083)}} \tag{4.5}$$

Now

$$\int_0^{t_r} 32000e^{-0.09t}e^{-0.1t}\,dt = \int_0^{t_r} 32000e^{-0.19t}\,dt = \left[\frac{-32000}{0.19}e^{-0.19t}\right]_0^{t_r} = 168421(1 - e^{-0.19t_r})$$

Therefore,

$$B(t_r) = \frac{168421(1 - e^{-0.19t_r}) - (15000 - 13600e^{-0.73t_r})e^{-0.1t_r}}{1 - \exp(-0.1(t_r + 0.083))} \tag{4.6}$$

Evaluating Equation 4.6 for various values of t_r gives Table 4.5. From the table it is clear that the benefits are maximized when replacement occurs at the end of the fourth year of operation.

4.3.4 FURTHER COMMENTS

In the example the time required to effect replacement has been included in the analysis. In practice, this time can usually be omitted since it is often small compared to the interval between replacements, and so it does not make any noticeable difference to the optimal replacement interval, whether it is included or not. However, all costs associated with the replacement time should be incorporated as part of the total cost of replacement.

It may seem unreasonable that we should sum the terms of the geometric progression to infinity. This, however, does make the calculations a little easier and gives an indication of the sort of interval we would expect to have between replacements. A dynamic programming approach assuming a finite planning horizon (White, 1969) can be applied equally well to capital replacement problems. The difficulty is, of course, to decide whether the more sophisticated analysis, which costs more to carry out, is likely to give a solution that is a significant improvement over the solution obtained by using a simpler model.

In practice, of course, new equipment comes on the market and we do not always replace equipment with identical equipment. Thus, as time goes on, we need to repeat our calculations using, when appropriate, new cost figures and so check whether it is necessary to modify the planned replacement interval.

The example of Section 4.5 gives an indication of how technological improvement can be incorporated into a model.

4.3.5 PROOF THAT OPTIMIZATION OVER A LONG PERIOD IS NOT EQUIVALENT TO OPTIMIZATION PER UNIT TIME WHEN DISCOUNTING IS INCLUDED

When dealing with long-term capital equipment replacement decisions, where the time value of money is taken into account, it is necessary to determine the replacement policy to maximize the performance measure (such as profit, cost, benefit, etc.) over a long period, and not to maximize performance per unit time, as is the case when dealing with the short-term decisions discussed in Chapter 2.

The basic problem is illustrated in Figure 4.13, where T is the period over which we wish to optimize t_r, the interval between replacements; $p(t_r)$ is the performance over one interval, which depends on the interval length t_r, assumed identical for each period of length t_r; P is the total discounted performance over period T, which we wish to optimize (we will assume we wish to maximize P); n is the number of replacement intervals in period $(0, T)$; and

$$\max(P) = \max[p(t_r) + p(t_r)e^{-it_r} + p(t_r)e^{-2it_r} + \ldots + p(t_r)e^{-(n-1)it_r}]$$

$$= \max\left[\frac{1-e^{-nit_r}}{1-e^{it_r}}\right]p(t_r) = \max\left[\frac{1-e^{-iT}}{1-e^{-it_r}}\right]p(t_r)$$

Since $(1-e^{-iT})$ is constant, therefore

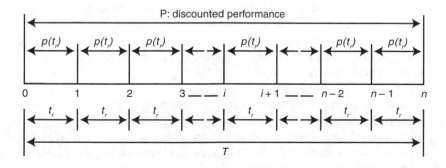

FIGURE 4.13 Optimization planning horizon.

$$\max(P) \equiv \max\left[\frac{p(t_r)}{1-e^{-it_r}}\right]$$

and not $\dfrac{p(t_r)}{t_r}$, which would be the result if discounting were neglected.

4.4 OPTIMAL REPLACEMENT INTERVAL FOR CAPITAL EQUIPMENT WHOSE PLANNED UTILIZATION PATTERN IS VARIABLE: MINIMIZATION OF TOTAL COST

4.4.1 Statement of Problem

Equipment when new is highly utilized, such as being on base load operations, but as it ages, its utilization decreases, perhaps due to being utilized only when there are peaks in demand for service. This class of problem is usually applicable to a fleet of equipment, such as a transportation fleet, where new buses may be highly utilized to meet base load demand, while older buses are used to meet peak demands, such as during the rush hour. In this case, when an item is replaced, the new one does not do the same work that the old one did, but is put onto base load operations, and the ones that were highly utilized are then less utilized as new units are put into service.

To establish the economic life of such equipment, it is necessary to examine the total cost associated with using the fleet to meet a specified demand. A model will be developed to establish the economic life of equipment operated in a varying utilization scenario such that the total costs to satisfy the demands of a fleet are minimized.

4.4.2 Construction of Model

1. A is the acquisition cost of the capital equipment.
2. $c(t)$ is the trend in operation and maintenance costs per unit time of equipment of age t; working age t can be measured in terms of utilization such as, in the case of vehicles, cumulative kilometers since new.
3. $y(x)$ is the utilization trend/period of the xth equipment to meet the annual demand; equipment is ranked from the newest (the 1st item) to the oldest (the Nth item).
4. S_i is the resale value of the equipment at the end of the ith period of operation, $i = 1, 2, ..., n$.
5. r is the discount factor.
6. n is the age in periods (such as years) of the equipment when replaced.
7. N is the fleet size.
8. $C(n)$ is the total discounted cost of operating, maintaining, and replacing the equipment (with identical equipment) over a long period with replacements occurring at intervals of n periods.

9. EAC(n) is the equivalent annual cost associated with replacing the equipment at age n periods.
10. The objective is to determine the optimal interval between replacements in order to minimize total discounted costs, $C(n)$, or equivalently, EAC(n). Note: EAC(n) = $C(n) \times$ CRF, and CRF = the interest rate used for discounting since $C(n)$ is calculated over an infinite period. However, it will be seen that in this case we simply obtain EAC(n) directly.

4.4.2.1 Consider a Replacement Cycle of n Years

In the steady state, the number or replacements per year will be N/n. Note: Most organizations wish to operate in a steady state; for example, if there is a fleet of 1000 buses and they are replaced on a 10-year cycle, then 1000/10 will be replaced each year.

Thus, the work undertaken by the newest N/n equipment will be

$$\int_0^{N/n} y(x)\,dx = D_1$$

and the cost of this will be obtained by considering the kilometers driven by one bus in its first year:

$$\int_0^{D_1/(N/n)} c(t)\,dt = C_1$$

In a similar way, the cost for other equipment in subsequent years can be obtained as C_2, C_3, C_4, ..., C_n.

The EAC associated with a replacement cycle of n years is then

$$\text{EAC}(n) = [A + C_1 + C_2 r^1 + C_3 r^2 + \ldots + C_n r^{n-1} - S_n r^n] \times \text{CRF}$$

$$= [A + \sum_{i=1}^{n} C_i r^{i-1} - S_n r^n] \times \text{CRF} \tag{4.7}$$

where $CRF = \dfrac{i(1+i)^n}{(1+i)^n - 1}$.

The optimal replacement age is that value of n that minimizes the right-hand side of Equation 4.7.

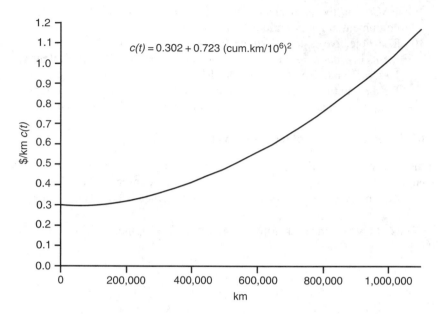

FIGURE 4.14 Operation and maintenance cost per kilometer.

4.4.3 NUMERICAL EXAMPLE

1. Let A = \$100,000, the price of a new vehicle.
2. Let $c(t) = 0.302 + 0.723$ (age of the vehicle in cumulative kilometers since new/10^6)^2. This is illustrated in Figure 4.14.
3. Let $y(x) = 80,000 - 40x$ km/year. This is illustrated in Figure 4.15.
4. The trend in resale values is provided in Table 4.6. Note: In this example it is assumed that there is essentially no resale value because it is specialized equipment — the resale values are really scrap values, with a 20-year-old asset being worth less than the others.
5. The interest rate for discounting is 6%.
6. The fleet size is 2000.

Using the above values in Equation 4.7 enables Figure 4.16 to be obtained, from which it is seen that the economic life of the equipment is 13 years with an associated EAC of \$28,230.

Sample calculation:

When $n = 20$ years the trend in utilization year by year is illustrated in Figure 4.17, from which the cash flows are obtained as depicted in Figure 4.18.

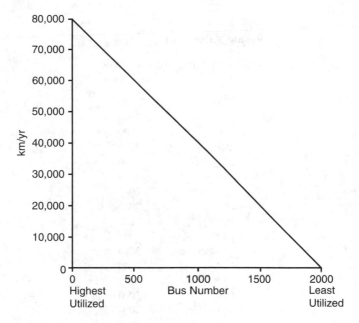

FIGURE 4.15 Equipment utilization trend.

1. The newest 100 buses will travel 7,800,000 km. Each bus travels 78,000 km.

$$\text{Cost per bus} = \$23,670/\text{year in first year}$$

2. Similarly, the next 100 buses each travel an average of 74,000 km/year in the second year.

$$\text{Cost per bus} = \$23,080 \text{ in second year}$$

$$\text{EAC} (20) = \$[100{,}000 + 23{,}670 + 23{,}080 \, (0.943) \ldots - 1000 \, (0.943)^{20}]$$

$$\times \frac{0.06(1+0.06)^{20}}{(1+0.06)^{20}-1} = \$29{,}973$$

4.4.4 FURTHER COMMENTS

The class of problem discussed in this section is typical of those found in many transport operations. For example:

TABLE 4.6
Resale Value Trend

Replacement Age (years)	Resale Value ($)
1	2000
2	2000
3	2000
4	2000
5	2000
6	2000
7	2000
8	2000
9	2000
10	2000
11	2000
12	2000
13	2000
14	2000
15	2000
16	2000
17	2000
18	2000
19	2000
20	1000

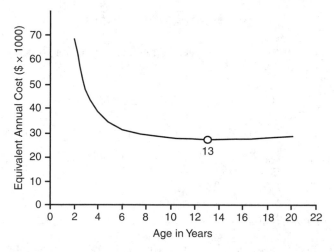

FIGURE 4.16 Equivalent annual cost trend.

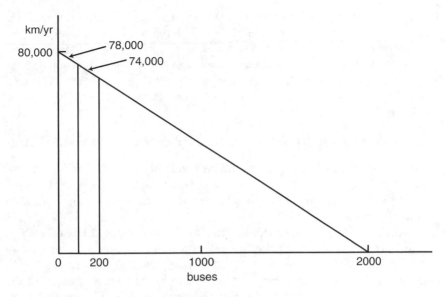

FIGURE 4.17 Calculating O&M cost per year.

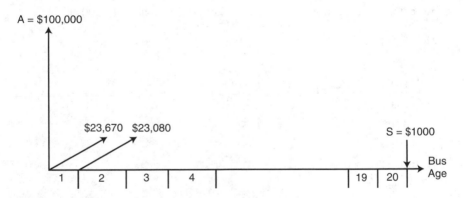

FIGURE 4.18 Cash flows when replacement age is 20 years.

- Mass transit fleets where new units are put into base load operation with older units used for peak morning and evening demands.
- Haulage fleets undertaking both long-distance and local deliveries — when new vehicles are used on long-haul routes initially, and then as they age are assigned to local delivery work.
- Stores with their own fleets of delivery vehicles, where there may be seasonal peaks in demand. The older vehicles in the fleet are retained to meet these predictable demands. Because of this unequal utilization, it is necessary to evaluate the economic life of the vehicle by viewing the fleet as a whole, rather than focusing on the individual vehicle.

In a machine shop with a group of similar machine tools, when a new machine tool is acquired, it may become the busiest, with the others being relegated and the oldest ones being disposed of. The same strategy may be applied to electrical generating stations. When a new station comes on line, it joins the group on base load operations, with the other coming into service to meet peak demands, such as morning and evening. And at some point one of the older stations will be decommissioned.

4.4.5 An Application: Establishing the Economic Life of a Fleet of Buses

A local authority wished to establish the economic life of its fleet of 54 conventional buses. The purchase price of a bus was $450,000; the trend in resale values was estimated and the interest rate for discounting was also known. The utilization trend is shown in Figure 4.19.

Using Equation 4.7, the economic life was calculated to be 13 years with an EAC of about $120,000. The practice in place within the transit authority was to replace a bus when it became 18 years old. Changing to a replacement age of 13 years provided a useful economic benefit. It also would have the benefit of the transit authority being seen to operate a rather new fleet of buses compared to the previous practice, but this intangible benefit is not incorporated in the model used.

A similar study conducted for the fleet of 2000 buses in Montreal, Canada, is presented in detail in Appendix 19 of Campbell and Jardine (2001).

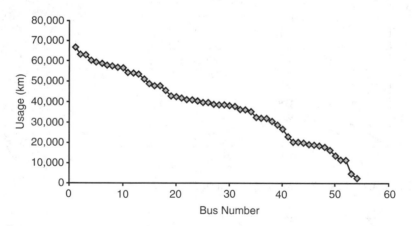

FIGURE 4.19 Bus utilization trend.

4.5 OPTIMAL REPLACEMENT POLICY FOR CAPITAL EQUIPMENT TAKING INTO ACCOUNT TECHNOLOGICAL IMPROVEMENT: FINITE PLANNING HORIZON

4.5.1 Statement of Problem

When determining a replacement policy, there may be on the market equipment that is in some way a technological improvement of the equipment currently used.

For example, maintenance and operating costs may be lower, throughput may be greater, quality of output may be better, and so on. The problem discussed in this section is how to determine when, if at all, to take advantage of the technologically superior equipment.

It is assumed that there is a fixed period from now during which equipment will be required, and if replacements are made using new equipment, then this equipment will remain in use until the end of the fixed period. The objective is to determine when to make the replacements, if at all, in order to minimize total discounted costs of operation, maintenance, and replacement over the planning horizon.

4.5.2 CONSTRUCTION OF MODEL

1. n is the number of operating periods during which equipment will be required.
2. $C_{p,i}$ is the operation and maintenance costs of the present equipment in the ith period from now, payable at time i, $i = 1, 2, \ldots, n$.
3. $S_{p,i}$ is the resale value of the present equipment at the end of the ith period from now, $i = 1, 2, \ldots, n$.
4. A is the acquisition cost of the technologically superior equipment.
5. $C_{t,j}$ is the operation and maintenance costs of the technologically superior equipment in the jth period after its installation and payable at time j, $j = 1, 2, \ldots, n$.
6. $S_{t,j}$ is the resale value of the technologically superior equipment at the end of its jth period of operation, $j = 0, 1, 2, \ldots, n$. ($j = 0$ is included so that we can then define $S_{t,0} = A$. This then enables Ar^0 in the model [see Equation 4.8] to be cancelled if no change is made.)
   Note: It is assumed that if a replacement is to be made at all, then it is with the technologically superior equipment. This is not unreasonable since it may be that the equipment currently in use is no longer on the market.
7. r is the discount factor.
8. The objective is to determine that value of T at which replacement should take place with the new equipment, $T = 0, 1, 2, \ldots, n$. The policy is illustrated in Figure 4.20.

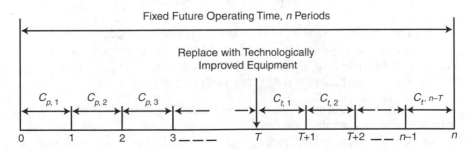

FIGURE 4.20 Technological improvement: finite planning horizon.

The total discounted cost over n periods, with replacement occurring at the end of the Tth period, is

$C(T)$ = discounted maintenance costs for present equipment over period $(0, T)$
    + discounted maintenance costs for technologically superior equipment over period (T, n)
    + discounted acquisition cost of new equipment
    − discounted resale value of present equipment at end of Tth period
    − discounted resale value of technologically superior equipment at end

$$\text{of } n^{th} \text{ period} = (C_{p,1}r^1 + C_{p,2}r^2 + C_{p,3}r^3 + \ldots + C_{p,T}r^T)$$

$$+ (C_{t,1}r^{T+1} + C_{t,2}r^{T+2} + \ldots + C_{t,n-T}r^n)$$

$$+ Ar^T - (S_{p,T}r^T + S_{t,n-T}r^n)$$

Therefore,

$$C(T) = \sum_{i=1}^{T} C_{p,i}r^i + \sum_{j=1}^{n-T} C_{t,j}r^{T+j} + Ar^T - (S_{p,T}r^T + S_{t,n-T}r^n) \qquad (4.8)$$

This is a model of the problem relating replacement time T to total discounted costs $C(T)$.

4.5.3 NUMERICAL EXAMPLE

1. The number of operating periods still to go, $n = 6$.
2. The estimated operation and maintenance costs $C_{p,i}$ over the next six periods for the present equipment are shown in Table 4.7.
3. The estimated trend in resale values of the present equipment payable at the end of the period is shown in Table 4.8.
4. The acquisition cost of the technologically superior equipment is $A =$ $10,000$.
5. The estimated operations and maintenance costs $C_{t,j}$ over the next six periods of the technologically superior equipment are shown in Table 4.9.
6. The estimated trend in resale value of the technologically improved equipment, payable at the end of its jth period, of operation $S_{t,j}$ is shown in Table 4.10.
7. The discount factor $r = 0.9$.

Evaluation of Equation 4.8 for different values of T gives Table 4.11, from which it is seen that the total costs are minimized when $T = 0$; that is, the technologically improved equipment should be installed now and used over the next six periods of operation.

TABLE 4.7
Trend in O&M Costs: Present Equipment

Period (i)	1	2	3	4	5	6
O&M Costs, $C_{p,i}$ ($)	5000	6000	7000	7500	8000	8500

TABLE 4.8
Trend in Resale Values: Present Equipment

Period (i)	0 (i.e., now)	1	2	3	4	5	6
Resale Value, $S_{p,i}$ ($)	3000	2000	1000	500	500	500	500

TABLE 4.9
Trend in O&M Costs: Technologically Improved Equipment

Period (j)	1	2	3	4	5	6
O&M Cost, $C_{t,j}$ ($)	100	200	500	750	1000	1200

TABLE 4.10
Trend in Resale Values: Technologically Improved Equipment

Period (j)	0	1	2	3	4	5	6
Resale Value, $S_{t,j}$ ($)	10,000	8000	7000	6000	5000	4500	4000

TABLE 4.11
Total Discounted Cost over the Planning Horizon

Replacement Time, T	0	1	2	3	4	5	6
Total Discounted Costs, $C(T)$	7211	10,836	14,891	18,649	22,062	25,519	28,359

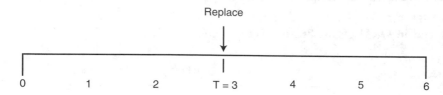

FIGURE 4.21 Sample calculation.

Note that if the minimum total cost occurs at $T = n$ (6 in this example), this would mean that no replacement would take place and the present equipment would be used for the remaining n periods of operation. If the minimum value of $C(T)$ occurs for a value of T between 0 and n, then the replacement should occur using the technologically improved equipment at the end of the Tth period.

Sample calculation (Figure 4.21):

When $T = 3$,

$$C(3) = C_{p,1}r^1 + C_{p,2}r^2 + C_{p,3}r^3 + Ar^3 + C_{t,1}r^4 + C_{t,2}r^5 + C_{t,3}r^6 - (S_{p,3}r^3 + S_{t,3}r^6)$$

$$= 5000(0.9) + 6000(0.9)^2 + 7000(0.9)^3 + 10,000(0.9)^3$$

$$+ 100(0.9)^4 + 200(0.9)^5 + 500(0.9)^6 - [500(0.9)^3 + 6000(0.9)^6]$$

$$= \$18,649$$

4.5.4 FURTHER COMMENTS

The example in this section assumed that once the decision was taken to replace the old equipment with the technologically improved equipment, no further replacements were made. In some situations, the time during which equipment is required is sufficiently long to warrant further replacements. Assuming that we continue to use the technologically superior equipment, it is not difficult to determine its economic life. Such a problem is covered in Section 4.6.

In addition, when dealing with technologically superior equipment, consideration may need to be given to capacity improvement and the effect it may have on the planning horizon.

4.5.5 AN APPLICATION: REPLACING CURRENT MINING EQUIPMENT WITH A TECHNOLOGICALLY IMPROVED VERSION

In a mining company there was an expected future mine life of 8 years, that is, a fixed planning horizon. A fleet of current, highly expensive equipment called a shovel was in use, and under normal circumstances, they would be used throughout the life of the mine. However, a new technologically improved shovel came on the market and the decision had to me made: Should the current equipment be used for the remaining 8 years, or should there be a changeover to the technologically superior equipment?

The model described by Equation 4.8 was modified to fit the mining company's goal of optimizing the changeover decision such that profit over the remaining mine life was maximized. In addition, the following features were included in the model: expected rate of return, depreciation rate, investment tax credit, capital cost allowance (CCA), depreciation type, federal, provincial, and mining tax rates, inflation rates, unit purchase year and price, unit yearly total operations and maintenance costs, unit yearly total production, unit yearly salvage values, and proposed replacement unit data. Thus, it is seen that the model of Equation 4.8 was extensively modified, and this is often what will occur with models presented in this book. They can be the foundation on which to build a more realistic model for the problem under study.

On conclusion of the study the following comments were made. "The equipment replacement system can be used to do the following analyses: compare the productivity of individual units with a fleet, find the "lemon" in a fleet of equipment (that is, the poorest-performing asset), calculate the optimum year to replace a unit, and, very importantly, use sensitivity testing to see what effect the rate of return, taxes, production, and other factors have on replacement timing." Details of the study are provided in Buttimore and Lim (1981).

4.6 OPTIMAL REPLACEMENT POLICY FOR CAPITAL EQUIPMENT TAKING INTO ACCOUNT TECHNOLOGICAL IMPROVEMENT: INFINITE PLANNING HORIZON

4.6.1 STATEMENT OF PROBLEM

The statement of this replacement problem is virtually identical to that of Section 4.5.1, except that once the decision has been taken to replace with the technologically improved equipment, then this equipment will continue to be used and a replacement policy (periodic) will be required for it. It will be assumed that replacement will continue to be made with the technologically improved equipment. Again, we wish to determine the policy that minimizes total discounted costs of operation, maintenance, and replacement.

4.6.2 CONSTRUCTION OF MODEL

1. $C_{p,i}$, $S_{p,i}$, A, $C_{t,j}$, $S_{t,j}$, and r are as defined in Section 4.5.2.
2. The replacement policy is illustrated in Figure 4.22.

The total discounted cost over a long period, with replacement of the present equipment at the end of T periods of operation, followed by replacement of the technologically improved equipment at intervals of length n, is

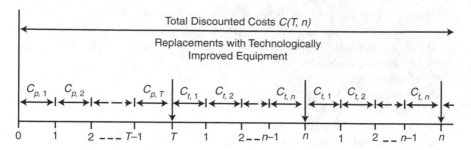

FIGURE 4.22 Technological improvement: infinite planning horizon.

$$C(T, n) = \text{costs over interval } (0, T) + \text{future costs}$$

$$\text{Costs over interval } (0, T) = \sum_{i=1}^{T} C_{p,i} r^i - S_{p,T} r^T + A r^T$$

Future costs, discounted to time T, can be obtained by the method described in Section 4.2.2 (Equation 4.1), where the economic life of equipment is calculated. We replace C_i by $C_{t,j}$ and use j as the counter for the summation to obtain

$$C(n) = \frac{\displaystyle\sum_{j=1}^{n} C_{t,j} r^j + r^n (A - S_n)}{1 - r^n} \tag{4.9}$$

Therefore, $C(n)$ discounted to time zero is $C(n) r^T$ and

$$C(T, n) = \sum_{i=1}^{T} C_{p,i} r^i - S_{p,T} r^T + A r^T + \left(\frac{\displaystyle\sum_{j=1}^{n} C_{t,j} r^j + r^n (A - S_n)}{1 - r^n} \right) r^T \tag{4.10}$$

This is a model of the problem relating changeover time to technologically improved equipment, T, and economic life of new equipment, n, to total discounted costs $C(T, n)$.

4.6.3 NUMERICAL EXAMPLE

Using the data of the example of Section 4.5.3, we can determine the economic life of the technologically improved equipment and the value of $C(n)$ in Equation 4.9. The data of Section 4.5.3 give Table 4.12, from which it is seen that the economic life is 5 years and the corresponding value of $C(n)$ is $11,920.

TABLE 4.12
Economic Life of Technologically Improved Equipment

Replacement Interval, n	1	2	3	4	5	6
Total Discounted Cost, $C(n)$	18,900	14,116	13,035	12,834	11,920	12,063

TABLE 4.13
O&M Cost of Present Equipment

Period, i	1	2	3
O&M Cost, $C_{p,i}$ ($)	1500	3000	4000

TABLE 4.14
Resale Values of Present Equipment

Period, i	0 (i.e., now)	1	2	3
Resale Value, $S_{p,i}$ ($)	2750	2500	1500	1000

TABLE 4.15
Optimal Replacement Time

Replacement Time, T	0 (i.e., now)	1	2	3
Total Discounted Cost, $C(T,5)$	19,170	18,828	20,120	20,946

Insertion of $C(5) = \$11,920$ and $A = \$10,000$ into Equation 4.10 gives

$$C(T,5) = \sum_{i=1}^{T} C_{p,i} r^i - S_{p,T} r^T + 10,000 r^T + 11,920 r^T$$

$$= \sum_{i=1}^{T} C_{p,i} r^i - S_{p,T} r^T + 21,920 r^T$$

(4.11)

Given the information of Table 4.13 and Table 4.14 for the operation and maintenance costs and resale prices for the present equipment, Table 4.15 can be obtained by evaluating values of $T = 0$, 1, 2, and 3 in Equation 4.11. Thus, it is seen that the present equipment should be used for one more year and then replaced with the technologically improved equipment, which should itself then be replaced at intervals of 5 years.

Sample calculation:

When $n = 3$, then Equation 4.9 becomes

$$C(3) = \frac{100(0.9) + 200(0.9)^2 + 500(0.9)^3 + (0.9)^3(10,000 - 6000)}{1 - 0.9^3}$$

$$= \$13,035$$

When $T = 2$, then Equation 4.11 becomes

$$C(2,5) = 1500(0.9) + 3000(0.9)^2 - 1500(0.9)^2 + 21,920(0.9)^2 = \$20,120$$

4.6.4 Further Comments

Of course, technological improvement is occurring continuously, and so perhaps we should cater to this in any model used for capital replacement. The real problem here is not construction of the model, but estimating the trends resulting from technological improvement, and this can be critical when establishing the economic time to replace an asset. On the assumption of exponential trends in benefits, operation, and replacement costs, Bellman and Dreyfus (1962) constructed a dynamic programming model that can be used to cater to technological improvement. Bellman and Dreyfus then extended the model to include the possibility of replacing old equipment with secondhand rather than new equipment.

4.6.5 An Application: Repair vs. Replace of a Front-End Loader

The model presented in Equation 4.10 can be used to examine the decision to repair (or rebuild) and compare this to the decision to completely replace an asset with a new one.

The decision problem is: Repair (or rebuild) today and then keep it to time T? Or, is it cheaper in the long run to replace today — thus the best value for T would be zero? Which alternative provides the minimum total cost? It is assumed in this case that if the equipment is repaired/rebuilt today, it can only be rebuilt once — there can be no further life extensions — and then it has to be replaced by new equipment. The cash flows associated with the alternative decisions are depicted in Figure 4.23.

The estimated cost for rebuilding the equipment was \$390,000, and the cost of acquiring new equipment, including costs associated with bringing it into service, was \$1.1 million.

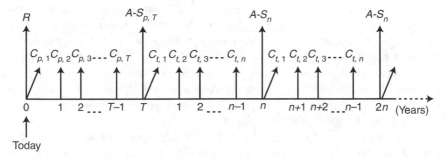

FIGURE 4.23 The repair vs. replace decision.

TABLE 4.16
Cost Data for Present Equipment

$C_{p,1}$ = \$138,592	$S_{p,0}$ = \$300,000
$C_{p,2}$ = \$238,033	$S_{p,1}$ = \$400,000
$C_{p,3}$ = \$282,033	$S_{p,2}$ = \$350,000
	$S_{p,3}$ = \$325,000

TABLE 4.17
The Optimal Repair vs. Replace Decision

	Changeover Time to New Loader, T			
	$T = 0$	$T = 1$	$T = 2$	$T = 3$
Overall EAC (\$)	449,074	456,744	444,334	435,237

The company made a commitment that if the equipment was rebuilt, it would only remain in service for 3 more years, and then a new asset would be purchased.

The estimated operation and maintenance costs for the present asset for the next 3 years, along with the expected trend in resale values, against a new one costing \$1.1 million, are given in Table 4.16. Note that the resale value today is \$300,000, but at the end of 1 year it is \$400,000. The reason for this increased value is that at the end of the next year of operation a rebuild would have taken place, costing \$390,000.

In addition to knowing the purchase price of new equipment, estimates were also obtained for the ongoing operation and maintenance costs of a new asset, as well as estimated resale values. Using Equation 4.1, it was concluded that the economic life of the new equipment was 11 years. The EAC was also obtained.

Using Equation 4.10, the results shown in Table 4.17 were obtained, from which it can be seen that the smallest EAC is at $T = 3$ years. However, it is also clear that there is very little difference in the EAC associated with replacing now ($T = 0$) and that associated with rebuilding and replacing in 3 years. This is a classic case where management would undoubtedly draw on additional insights before making a final repair/rebuild or replace decision. At the commencement of the study there were two distinct camps: those who believed the most economic decision was to rebuild the asset and those who were convinced that the best decision was to replace the asset immediately. Once the data were analyzed, it was clear that there really was no real economic difference between the two alternatives.

A further comment: in Chapter 2, Section 2.2.2, a general rule was presented that said the optimal replacement time is when the current rate of operating cost is equal to the average total cost per unit time. A similar rule can be made for the repair/rebuild vs. replace decision; that is, replace if the EAC for the next period is greater than the current EAC. In the engineering economic literature, see, for example, Park et al. (2000); this is often termed the challenger problem, since the new equipment is offering a challenge to the old equipment that is presently in use in that it is demanding to be used and asking for the old asset to be discarded. This rule requires that the trend in equipment O&M costs is monotonically nondecreasing; thus, there can be no decrease in next year's O&M costs compared to the current costs. In asset management, major maintenance actions are often taken in a year, knowing that there will be lower costs in subsequent years. If this is possible, then care must be taken before applying the rule mentioned above.

4.7 SOFTWARE FOR ECONOMIC LIFE OPTIMIZATION

4.7.1 INTRODUCTION

Rather than solve the mathematical models for capital equipment from first principles, software that has the models programmed in provide a very easy way to solve the models. Two such packages are PERDEC and AGE/CON (www.banak-inc.com). In this section, use will be made of the educational versions of these two packages, which can be downloaded freely from the publisher's Web site.

PERDEC (an acronym for Plant and Equipment Replacement Decisions) is geared for use by the fixed plant community; AGE/CON (based on the French term L'Age Économique) is designed for use by the fleet community.

Both have the same mathematical models in them since there is no difference mathematically as to whether one is establishing the economic life of a piece of fixed equipment (such as a machine) or a piece of mobile equipment (such as a vehicle). However, there are slight differences in vocabulary. For example, if PERDEC is used, in the opening screen where O&M costs are entered, the column is headed "machine(s)." If AGE/CON is used, in the opening screen where O&M costs are entered, the column is headed "vehicle(s)."

If PERDEC is used, when "Parameters" are selected and then "Constant annual utilization" is selected, the analysis will be described as dealing with

TABLE 4.18
Data Entry and AGE/CON Solution

AGE/CON - Main

File Edit View Parameters Help

Description Tractor

Number of Years 5 Acquisition Cost 85,000 Best Year 1

Parameters

Age of Vehicle(s)	O&M Cost	Resale Value	Resale Rate(%)	EAC
1 Year Old	29,352	60,000	70.5882	65,787
2 Years Old	45,246	40,000	47.0588	70,541
3 Years Old	52,626	25,000	29.4118	72,459
4 Years Old	53,324	20,000	23.5294	71,101
5 Years Old	42,363	15,000	17.6471	68,234

Copyright © 2001, Andrew Jardine   Licenced to Andrew Jardine Evaluation

"utilization of the machine." If AGE/CON is used, when "Parameters" are selected and then "Constant annual utilization" is selected, the analysis will be described as dealing with "utilization of the vehicle."

Similar differences can be spotted elsewhere in the software. Fixed equipment people do not like to see their equipment called a vehicle, and fleet people do not like to have their vehicles called machines.

4.7.2 USING PERDEC AND AGE/CON

We will use the data provided in Section 4.2.5, namely: purchase price = $85,000. The trend in O&M costs, resale values, and interest rate for discounting are given in Table 4.4.

Entering these values into AGE/CON, we get the screen dump of Table 4.18, from which it is seen that the economic life is 1 year with an associated EAC of $65,787. (The interest rate of 10% is entered after the parameter button is hit, and so is hidden in the screen dump.)

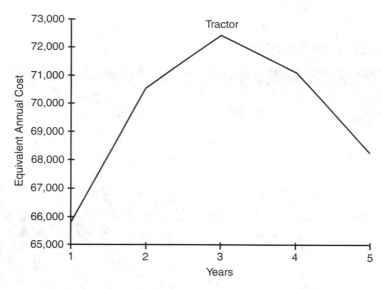

FIGURE 4.24 EAC vs. replacement age from AGE/CON.

The EAC graph is provided in Figure 4.24, showing very clearly that the economic life is 1 year.

4.7.3 FURTHER COMMENTS

This section has just dipped very briefly into software packages that can be used to optimize replacement of capital equipment. Other packages are available, including a life cycle cost worksheet from www.Barringer1.com.

One of the major benefits of using a software package is the ease with which sensitivity analyses can be undertaken.

REFERENCES

Bellman, R.E. and Dreyfus, R. (1962). *Applied Dynamic Programming*. Oxford: Oxford University Press.

Buttimore, B. and Lim, A. (1981). Noranda equipment replacement system. In *Applied Systems and Cybernetics*, Vol. II, G.E. Lasker, Ed. Oxford: Pergamon Press, pp. 1069–1073.

Campbell, J.D. and Jardine, A.K.S. (2001). *Maintenance Excellence: Optimizing Equipment Life-Cycle Decisions*. New York: Marcel Dekker.

Christer, A.H. and Waller, M.W. (1987). Tax-adjusted replacement models. *Journal of the Operational Research Society*, 993–1006.

Eilon, S., King, J.R., and Hutchinson, D.E. (1966). A study in equipment replacement. *Operational Research Quarterly*, 17.

Park, C., Pelot, R., Porteous, K., and Zuo, M.J. (2000). *Contemporary Engineering Economics: A Canadian Perspective*, 2nd ed. Pearson Education.

White, D.W. (1969). *Dynamic Programming*. Oliver and Boyd.

PROBLEMS

The following problems are to be solved using the mathematical models.

1. A new machine tool costs $5000. Its associated trends in operating costs and resale value are given in Table 4.19. Determine the optimal replacement interval to minimize total cost per year (assume no interest rate is applicable).
2. Assuming that the interest rate for discounting purposes is $i = 8\%$, compounded annually, determine the optimal replacement age for the machine tool data of problem 1.
3. Parks and recreation equipment used by Moose, Inc., cost $17,000. Its trends in operating cost and resale value are given in Table 4.20.

   Construct a mathematical model that can be used to determine the economic replacement age of the equipment.

TABLE 4.19
Trend in Machine Tool Costs

Year	Operating Costs ($)	Resale Value ($)
1	2500	3000
2	2750	1800
3	3025	1080
4	3330	650
5	3660	400
6	4025	400
7	4425	400
8	4850	400

TABLE 4.20
Trend in Land Cruiser Costs

Year	Operating Costs ($)	Resale Value ($)
1	1000	3000
2	1500	1000
3	2500	0
4	2500	0
5	5000	0
6	10,000	0
7	15,000	0

TABLE 4.21
NC Machine Cost Data

Acquisition cost of a new machine tool	$77,000
Operating cost of tool in its first year of life	$25,000
Operating cost of tool in its second year of life	$35,000
Operating cost of tool in its third year of life	$50,000
Operating cost of tool in its fourth year of life	$65,000
Scrap value of tool	$3000
Interest rate for discounting purposes = 15%	

TABLE 4.22
Bus Resale Values

Year	Resale Value
1	$75,000
2	$60,000
3	$30,000
4	$10,000
5	$5000

Given that the cost of capital is 10% per annum, what is the economic life? Show your calculations.

4. The operating costs of a numerically controlled machine tool seem to be becoming excessive, and it has been decided to analyze some data to determine the economic life of the tool. Given the data of Table 4.21, what is the economic life of the machine?

State clearly the economic life model used and the method of solution.

5. The acquisition cost of a bus is $100,000. The trend in operating costs can be given by the equation

$$\text{\$/km} = 0.5 + 5 \times 10^{-6}\, d$$

where d is the number of kilometers traveled from new.

A bus travels an average of 80,000 km per year, and it does not depend on the bus's age. The trend in resale value for a bus in its first 5 years of life is given in Table 4.22.

a. Making appropriate assumptions, construct a model that could be used to determine the economic life of a bus.

b. Using the model constructed in (a), along with the above data, and taking i to be 10% per annum, what is the economic life of a bus?

TABLE 4.23
Side-Loader Cost Data

Year	Average Operating and Maintenance Cost ($/year)	Resale Value at End of Year ($)
1	16,000	100,000
2	28,000	60,000
3	46,000	50,000
4	70,000	20,000

TABLE 4.24
Rental Automobile O&M Data

	Average O&M Costs
4 years ago (first year of vehicle life)	$1800
3 years ago (second year of vehicle life)	$3400
2 years ago (third year of vehicle life)	$6800
Last year (fourth year of vehicle life)	$13,700

The following are to be solved using the educational versions of the AGE/CON or PERDEC software.

6. Canmade Ltd. wants to determine the optimal replacement age for its turret side-loaders to minimize total discounted costs. Historical data analysis has produced the information (all costs in present-day dollars) contained in Table 4.23.

   The cost of a new turret side-loader is $150,000, and the interest rate for discounting purposes is 12% per annum.

   Find the optimal replacement age for the side-loaders.

7. An automobile rental company has kept records on a particular type of vehicle. Historical data are therefore available for 12 of these automobiles that came into service on the same date, 4 years ago. The O&M costs are provided in Table 4.24.

   The purchase price today for a new automobile is $32,000, and current trade-in values of this type of vehicle are given in Table 4.25.

   Assume that the interest rate for discounting purposes is 16% per annum and that the average inflation rate during the last 4 years has been 9% per annum.

   Find the optimal replacement age for these automobiles.

8. A new dump truck costs $45,000, and its associated trends in operating cost and resale value are given in Table 4.26.

TABLE 4.25
Rental Automobile Resale Values

1-year-old vehicle	$19,000
2-year-old vehicle	$12,000
3-year-old vehicle	$8000
4-year-old vehicle	$4500

TABLE 4.26
Dump Truck Costs

Year	Operating Costs ($)	Resale Value ($)
1	6000	28,000
2	10,000	18,000
3	15,000	10,000
4	22,000	5000

TABLE 4.27
Minibus O&M Costs

	Average O&M Costs
4 years ago (first year of bus life)	$10,000
3 years ago (second year of bus life)	$13,500
2 years ago (third year of bus life)	$17,000
Last year (fourth year of bus life)	$20,500

Find the optimal replacement age for a dump truck. (Assume no interest rate is applicable.)

9. Assuming the interest rate for discounting purposes is $i = 10\%$, compounded annually, determine the optimal replacement age for the dump truck data in question 8.

10. Repeat question 9, this time basing the economic life on after-tax dollars. Assume that capital cost allowance (CCA) is equal to 30% and that corporation tax rate is 50%.

11. A big sports club has its own fleet of eight minibuses. The club has kept records for these eight buses, which all came into service on the same date 4 years ago. The O&M costs are given in Table 4.27.

The purchase price today for a new minibus is $70,000, and the current trade-in values for this kind of minibus are given in Table 4.28.

TABLE 4.28
Minibus Trade-In Values

1-year-old bus	$43,000
2-year-old bus	$29,000
3-year-old bus	$20,000
4-year-old bus	$14,000

TABLE 4.29
Minibus Utilization Pattern

Year	Utilization (miles)
4 years ago	15,000
3 years ago	13,000
2 years ago	10,000
Last year	7000

Due to a general decline in the sports club economy, traveling activities have become less popular in the last 4 years. As a result, the average utilization of the minibuses has not been constant but is as depicted in Table 4.29.

Assume that the interest rate for discounting purposes is 12% per annum and that the average inflation in the last 4 years has been 7% per annum.

Assume a future average annual utilization of 10,000 miles and find the optimal replacement age for these minibuses and the associated EAC. Check the effect on the economic life and the associated EAC value of these minibuses if they are used for an average of only 8000 miles/year.

12. Mosal Ltd. wants to find the optimal replacement age for its forklift trucks (FLTs). Historical data collected over the last 4 years have produced the information in Table 4.30. (Assume the whole fleet of FLTs was new 4 years ago.)

The current trade-in values for the FLTs are given in Table 4.31.

The price of the new forklift truck is $51,000, and the interest rate for discounting purposes is 14% per annum.

Find the optimal replacement age and the associated EAC for the forklift trucks.

13. Repeat question 12, this time assuming that no resale values are applicable, but only a scrap value of $8500.

TABLE 4.30
Forklift Truck Data

Year	O&M Inflation (%)	Average O&M Cost ($)
4 years ago	6	5100
3 years ago	7	10,300
2 years ago	8	17,100
Last year	7	29,000

TABLE 4.31
Forklift Truck Trade-In Values

1-year-old truck	$34,000
2-year-old truck	$24,000
3-year-old truck	$17,000
4-year-old truck	$11,000

TABLE 4.32
Baggage-Handling Truck Data

Year	Average O&M Costs ($)	Average Utilization of Trucks (miles/year)
4 years ago	6200	4000
3 years ago	7700	8000
2 years ago	13,700	7000
Last year	22,400	7500

14. An airline operator has its own fleet of 30 airline baggage-handling trucks. The operator has kept good records for the last 4 years, so the data of Table 4.32 are available.

The purchase price today for a new truck is $38,000. The current trade-in values for a truck are given in Table 4.33.

Assume that the interest rate for discounting purposes is 14% per annum and that the average inflation rate during the last 4 years has been 8% per annum.

The airline operator is expecting a future average annual utilization of 7000 miles/year for its trucks.

Find the optimal replacement age for these trucks.

Check the result on the answer when the trend line for the O&M cost is changed (i.e., the degree of the fitted polynomial is changed).

TABLE 4.33
Baggage-Handling Truck
Trade-In Values

1-year-old truck	$25,000
2-year-old truck	$17,000
3-year-old truck	$10,000
4-year-old truck	$4500

TABLE 4.34
Vehicle Annual Utilization

Vehicle 1 goes	23,300 miles/year
Vehicle 2 goes	19,234 miles/year
Vehicle 3 goes	15,876 miles/year
Vehicle 4 goes	15,134 miles/year
Vehicle 5 goes	12,689 miles/year
Vehicle 6 goes	8756 miles/year
Vehicle 7 goes	3422 miles/year
Vehicle 8 goes	1589 miles/year

15. A company keeps a fleet of eight delivery vehicles to carry its products to the customers. The company runs a policy of utilizing its newest vehicles during normal demand periods, and using the older ones to meet peak demands.

Suppose the whole fleet travels 100,000 miles/year, and these miles are distributed among the eight vehicles as shown in Table 4.34.

This trend line can be described by the general equation

$$Y = a + bX$$

where Y is the miles per year figure and X is the vehicle number — vehicle 1 is the most utilized and vehicle 8 is the least utilized.

Using the actual figures given above, it can be shown (using a simple software package or simply by plotting the data) that the equation in our case would read

$$Y = 26,152 - 3034X$$

A plot of these figures is illustrated in Figure 4.25.

Now we must figure out the trend for operations and maintenance costs.

The O&M costs per mile are $0.48 for vehicle 8.

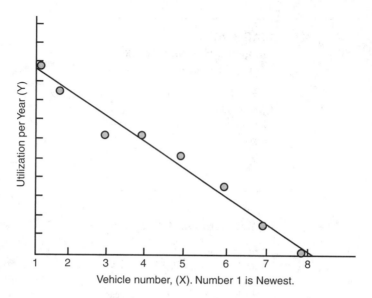

FIGURE 4.25 Annual utilization trend.

For vehicle 1 (our newest vehicle and the one we utilize the most), the following information is available:

Miles traveled last year   23,300
O&M costs last year $3150
Cumulative miles on the odometer to the mid-point of last year   32,000

Thus, the O&M cost per mile is $0.14 for vehicle 1. Then we do the same for all eight vehicles. Vehicle 8 (the oldest vehicle and the one we utilize the least) may look like this:

Miles traveled last year (already given)   1589
O&M costs last year $765
Cumulative miles on the odometer to the mid-point of last year 120,000

It will be necessary to fit a trend line through these points. It will look something like Figure 4.26.

Each vehicle has a "dot" — vehicles 1 and 8 are identified on the graph in the figure. The straight line is the trend line we have fitted to the dots.

The equation we get this time is

$$Z = 0.0164 + 0.00000394T$$

where Z is dollars per mile and T is cumulative miles traveled.

The two trend lines obtained are both used as input to AGE/CON.

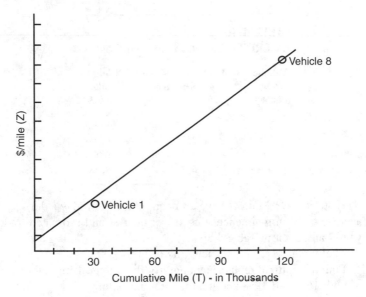

FIGURE 4.26 Trend in O&M costs.

TABLE 4.35
Delivery Vehicle Trade-In Values

1-year-old vehicle	$28,000
2-year-old vehicle	$20,000
3-year-old vehicle	$13,000
4-year-old vehicle	$6000

Note: In both cases we found that a straight (linear) relationship existed for $Y(X)$ and $Z(T)$ so that the fitted lines read $Y = a + bX$ and $Z = c + dT$. Often, a polynomial equation will give a better fit to a particular set of data. These polynomial equations can be generated by using a software package.

To go on with our problem, we need some more information: Assume that a new delivery vehicle costs $40,000. The resale values for this particular type of vehicle are given in Table 4.35. The interest rate for discounting purposes is 13% per annum. Find the optimal replacement age for a delivery vehicle.

16. AMERTRUCK keeps good records, and it knows the quarterly costs associated with maintaining its vehicles since they entered the fleet. It has a fleet of 11 heavily utilized 96,000-lb tractors, each doing approximately 32,000 miles/quarter.

TABLE 4.36
AMERITRUCK Maintenance Costs

Vehicle Age (quarter)	Maintenance Costs per Quarter ($)	Trade-In Value ($)
1	1435	57,500
2	2334	40,000
3	3974	30,000
4	6176	19,000

Historical O&M costs have all been inflated to today's dollars. The average trend in maintenance costs per quarter and estimated trade-in values are given in Table 4.36.

The purchase price of a new tractor is $70,000.

Find the optimal replacement age for the tractor using interest rates of both 16% and 19% per annum (see hint below).

Hint: Since we are working in quarters, we must convert the annual interest rate to an equivalent quarterly rate. This is not done by simply dividing the annual interest rate by four, but as follows: If the annual rate is 16%, then $(1+i)^4 = 1.16$, where i = interest rate/quarter that is equivalent to 16% per annum. Therefore, $i = 0.0378$, or 3.78%.

Similarly, when the annual interest rate is 19%, the equivalent quarterly rate is determined to be 4.44% per quarter.

5 Maintenance Resource Requirements

There is one and only one social responsibility of business — to use its resources and engage in activities designed to increase its profits so long as it stays within the rules of the game.

Milton Friedman

5.1 INTRODUCTION

The goal of this chapter is to present models and tools that can be used to determine optimal resource requirements to optimize physical asset management resource decisions. In the context of the framework of the decision areas addressed in this book, we are addressing column 4 of the framework, as highlighted in Figure 5.1.

The two interrelated problem areas, concerning what type of maintenance organization should be created, that will be considered in this chapter are:

1. Determination of what facilities (e.g., staffing and equipment) there should be within an organization
2. Determination of how these facilities should be used, taking into account the possible use of subcontractors (i.e., outside resources)

5.1.1 THE FACILITIES FOR MAINTENANCE WITHIN AN ORGANIZATION

Within an organization there are generally some maintenance facilities available, such as workshops, stores, and tradespeople. In addition, there is usually some form of arrangement between the organization and the contractors who are capable of performing some or all of the maintenance work required by the organization.

The problem is to determine the best composition of facilities for maintenance. An increase in the range of maintenance equipment, such as lathes, increases the capital tied up in plant and buildings and requires an increase in staffing. Increases in the in-plant facilities, however, will reduce the need to use outside resources such as general engineering contractors. In this case, a balance is required between costs associated with using in-plant facilities and costs of using outside resources. A difficult costing problem arises since not only does the cost charged by the outside resource have to be considered, but also the cost associated with loss of control of the maintenance work by management. Also,

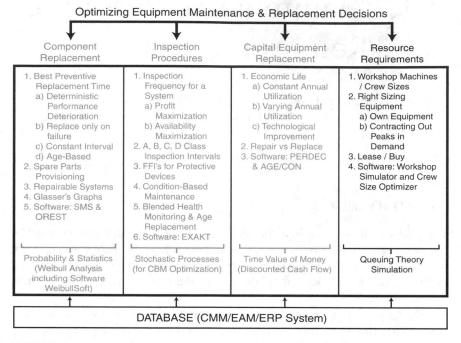

FIGURE 5.1 Resource requirements.

by using outside resources, there is the possibility of greater downtime occurring on production equipment, and so a cost must be associated with this downtime.

Also within this area, there is the problem of determining the size of the maintenance crew. The major conflicts arising here are that:

1. As crew size increases, so does its cost.
2. As crew size increases, the time that machines are idle, waiting for a member of the maintenance crew, decreases.
3. Downtime may be reduced since larger crews can be used to repair equipment.

5.1.2　The Combined Use of the Facilities within an Organization and Outside Resources

Maintenance work can be performed by either company personnel or contractors, on company premises or at contractors' premises. Just which of these alternatives are invoked at any particular time will depend on:

1. The nature of the maintenance work required
2. The maintenance facilities available within the company
3. The workload of these facilities
4. The costs associated with the various alternatives

It should be noted that these alternatives are not mutually exclusive since maintenance work (e.g., the repair of a piece of equipment or a complete production line) can be done by cooperation between the company's facilities and outside resources.

5.2 QUEUING THEORY PRELIMINARIES

If there are not sufficient resources available within an organization for undertaking the required maintenance workload, this will be very visible — such as there being a queue of jobs to process (there will be a large backlog) and operations being quite unhappy with the service provided by maintenance. There is a branch of mathematics known as queuing theory (or waiting-line theory) that deals with problems of congestion, where "customers" arrive at a service facility, perhaps wait in a queue, are served by "servers," and then leave the service facility. These customers may be machines requiring repair and waiting for a maintenance crew, or jobs waiting to be processed on a workshop machine. Thus, queuing theory is very valuable when tackling problems where there is a bottleneck (queue) in a system and one is exploring the potential benefit of adding more resources to deliver an improved service.

The problem of Section 5.3 uses results obtained from the mathematical theory of queues (or waiting-line theory), so we will first give a brief introduction to the relevant aspects of this theory.

For a given service facility (e.g., workshop size, maintenance crew size), what is the average time that a job has to wait in a queue?

For a given service facility, what is the average number of jobs in the system at any one time?

For a given service facility and given pattern of workload, what is the average idle time of the facility?

For a given service facility, what is the probability of a waiting time greater than t?

For a given service facility, what is the probability of one of the servers in the facility being idle?

Once information such as the above is obtained, it may then be possible to identify the optimal size of the service facility to minimize the total cost of the service facility and downtime incurred due to jobs waiting in a queue for service. These basic conflicts are illustrated in Figure 5.2.

5.2.1 QUEUING SYSTEMS

Figure 5.3 and Figure 5.4 depict the usual queuing systems we deal with. Figure 5.3 is the situation where there is a single-server facility (i.e., single channel) and only one customer can be served at any time. All incoming jobs join a queue, unless the service facility is idle, and eventually depart from the system.

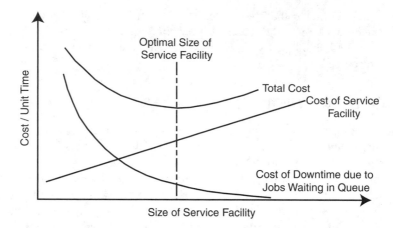

FIGURE 5.2 Optimizing the service facility size.

FIGURE 5.3 Single-channel queuing system.

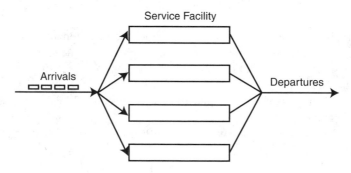

FIGURE 5.4 Multichannel queuing system.

Figure 5.4 is a multichannel system where customers join a queue and then go from the queue to the first service facility that becomes vacant.

Before analysis of a queuing system can be undertaken, the following information must be obtained:

1. The arrival pattern of customers. In this chapter the arrival pattern will be assumed to be random; the interval between the arrival of jobs at the service facility will be negative exponentially distributed. Thus, we are dealing with a Poisson process where the number of arrivals in a specified period is distributed according to the Poisson distribution. See Section A1.3.2 of Appendix 1.

2. The service pattern of the facility. In this chapter the service distribution is assumed to be negative exponential; the time taken to repair a job in the service facility is negative exponentially distributed.
3. The priority rules. In this chapter the priority rule is that customers are served (or begin to receive service) in the order of their arrival.

In practice, the assumptions made in 1 to 3 are often acceptable, although other patterns of arrival or service, or priority rule, may be appropriate. When this is the case, then the general results of queuing theory used in this chapter may not be applicable, and the reader will have to seek guidance in some of the standard references to queuing theory, such as Gross and Harris (1997) or Cox and Smith (1961).

When dealing with complex queuing situations, it is often the case that analytical solutions cannot be obtained, and in this case, we may resort to simulation. This will be covered in the problem of Section 5.4.

5.2.2 QUEUING THEORY RESULTS

5.2.2.1 Single-Channel Queuing System

This type of queuing system has the following characteristics:

Poisson arrivals, negative exponential service, customers served in order of their arrival
λ, mean arrival rate of jobs per unit time
$1/\lambda$, mean time between arrivals
μ, mean service rate of jobs per unit time (if serving facility is kept busy).

Then we can calculate the following statistics, which apply in the steady state, that is, when the system has settled down:

Mean waiting time of a job in the system, $W_s = 1/(\mu - \lambda)$
Mean time one job waits in a queue, $W_q = \rho/(\mu - \lambda)$, where ρ is termed the traffic intensity, λ/μ

Note that to ensure that an infinite queue does not build up, ρ must always be less than 1. The above results for W_s and W_q depend on this assumption.

5.2.2.2 Multichannel Queuing Systems

Although formulas are available for waiting times, and so forth, in certain multichannel systems, with particular arrival and service patterns, they are beyond the scope of this book. However, tables and charts are available that enable us to obtain directly the quantities we need. Such tables include those of Peck and Hazelwood (1958). The chart of Figure 5.5, which is taken from Wilkinson (1953), is used to determine the mean waiting time of a job in the system. A similar chart appears in Morse (1963), as do charts of other queuing statistics.

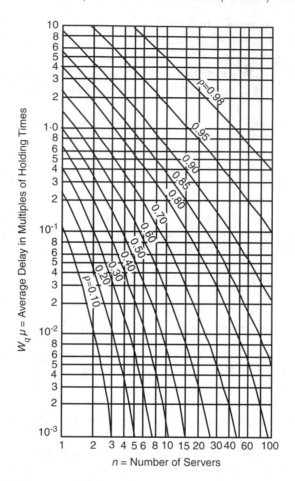

FIGURE 5.5 Wilkinson queuing chart. (Reprinted with permission of John Wiley & Sons, Inc.)

5.3 OPTIMAL NUMBER OF WORKSHOP MACHINES TO MEET A FLUCTUATING WORKLOAD

5.3.1 STATEMENT OF PROBLEM

From time to time jobs requiring the use of workshop machines (e.g., lathes) are sent from various production facilities within an organization to the maintenance workshop. Depending on the workload of the workshop, these jobs will be returned to production after some time has elapsed. The problem is to determine the optimal number of machines that minimizes the total cost of the system. This cost has two components: the cost of the workshop facilities and the cost of downtime incurred due to jobs waiting in the workshop queue and then being repaired.

5.3.2 CONSTRUCTION OF MODEL

1. The arrival rate of jobs to the workshop requiring work on a lathe is Poisson distributed with arrival rate λ.
2. The service time a job requires on a lathe is negative exponentially distributed with mean $1/\mu$.
3. The downtime cost per unit time for a job waiting in the system (i.e., being served or in the queue) is C_d.
4. The cost of operation per unit time for one lathe (either operating or idle) is C_l.
5. The objective is to determine the optimal number of lathes n to minimize the total cost per unit time $C(n)$ of the system.

$C(n)$ = cost per unit time of the lathes + downtime cost per unit time due to jobs being in the system

Cost per unit time of the lathes
    = number of lathes $\times$ cost per unit time per lathe = nC_l

Downtime cost per unit time of jobs being in the system
    = average waiting time in the system per job
        $\times$ arrival rate of jobs in the system per unit time
        $\times$ downtime cost per unit time per job = $W_s \lambda C_d$

where W_s = mean waiting time of a job in the system. Hence,

$$C(n) = nC_l + W_s \lambda C_d \qquad (5.1)$$

This is a model of the problem relating number of machines n to total cost $C(n)$.

The problem is depicted in Figure 5.6.

5.3.3 NUMERICAL EXAMPLE

Letting $\lambda = 30$ jobs/week, $\mu = 5.5$ jobs/week (for one lathe), $C_d = \$500$/week, $C_l = \$200$/week, Equation 5.1 can be evaluated for different numbers of lathes to give the results shown in Table 5.1. Thus, it is seen that the optimal number of lathes to minimize total cost per week is 8.

Figure 5.7 illustrates the underlying pattern of downtime and lathe costs that, when added together, give the total costs of Table 5.1.

It is also interesting to plot Figure 5.8, which gives the average idle time and busy time per week for each lathe for different numbers of lathes. Note that when $n = 8$, the optimal number from a total cost viewpoint, the average idle time of a lathe is 32%; that is, utilization is 68%. So often the comment is made that a high utilization for equipment is required, and only then is it being operated

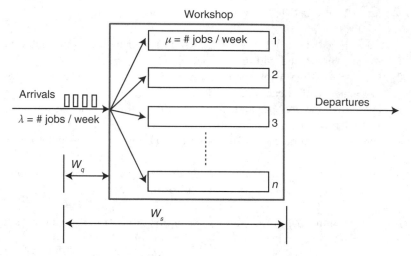

FIGURE 5.6 Workshop machine system.

TABLE 5.1
Optimal Number of Lathes

Number of Lathes, n	Mean Wait of a Job in the System, W_s (weeks)	Total Cost per Week, $C(n)$ ($)
6	0.437	7755
7	0.237	4955
8	0.198	4570
9	0.189	4635
10	0.185	4775
11	0.183	4945
12	0.182	5130

efficiently. In some cases this will be so, but we see from this example that if the utilization of a lathe were increased from 68 to 91% (which would occur when $n = 6$), the total cost per week would increase from $4570 to $7750. Again, the point can be made that we must be clear in our mind about what objective we are trying to achieve in our maintenance decisions.

Sample calculations:

When $n = 1$ to 5, then $\rho = \dfrac{\lambda}{n\mu}$, the traffic intensity, is greater than 1. Thus,

an infinite queue will eventually build up since work is arriving faster than

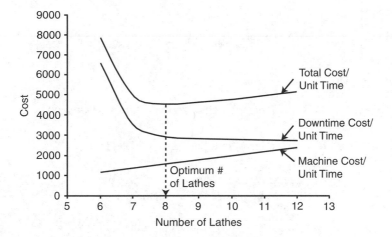

FIGURE 5.7 Optimal number of lathes.

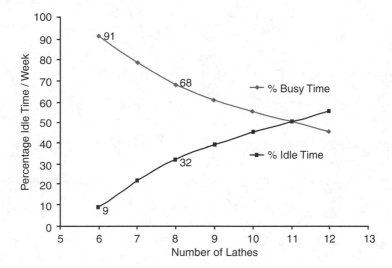

FIGURE 5.8 Lathe utilization statistics.

it can be processed, and so we consider cases of n at least equal to 6. (Note that the formulas apply to the steady state. In practice, an infinite queue cannot be formed.)

From Figure 5.5, when $n = 6$, $\rho = 0.91$, then $W_q\mu = 1.4$.

Therefore,

Mean waiting time in a queue $W_q = 1.4/5.5 = 0.255$ week

Therefore,

$$W_s = W_q + \text{mean service time} = 0.255 + 0.182 = 0.437 \text{ week}$$

From Equation 5.1,

$$C(6) = 6 \times \$200 + 0.437 \times 30 \times \$500 = 1200 + 6555 = \$7755$$

To calculate the average busy time per week for one lathe:

Average busy time per week = average number
of jobs to be processed on a lathe per week

$$\times \text{average time of one job on a lathe} = \frac{\lambda}{n} \times \frac{1}{\mu}$$

Therefore,

$$\text{Average idle time per week} = 1 - \frac{\lambda}{n\mu}$$

When $n = 6$, $\lambda = 30$, $\mu = 5.5$, then

$$\text{Average busy time per week for one lathe} = \frac{30}{6 \times 5.5} = 0.91$$

$$\text{Average idle time per week for one lathe} = 1 - 0.91 = 0.09$$

Note that ρ, the traffic intensity, is equivalent to the average busy time per week.

5.3.4 FURTHER COMMENTS

The goal of the model in this section was to optimize the number of servers (lathes) such that total cost was minimized. In practice, in many such problems the cost of the resource is not too difficult to quantify, such as the cost of additional lathes, but a difficult costing problem may arise when associating a cost with the improvement of service with an increased level of a resource. In this section, the cost benefit of reducing the waiting time of jobs in the workshop system is evaluated as more lathes are added. Because of this difficulty, the analysis sometimes stops at identifying the quality of service, such as average throughput rate, for a given level of the resource. The final resource level decision is then made by management through selecting an appropriate compromise between resource cost and service level provided.

The method of tackling the lathe problem of this section is the same as that which could be adopted to determine the optimal size of a maintenance crew. In that case, the number of tradespeople in the crew corresponds to the number of machines in the lathe group. One report of such a study is that of Carruthers et al. (1970).

In the problem of this section it was assumed that all the machines were the same, and any machine could be used equally well for any job requiring lathe work. This may not be the case. For example, within a group of lathes there may be small, medium, and large lathes. Certain incoming jobs may be done equally well on any of the lathes, but others may only be processed on, say, a large lathe. This sort of problem will be discussed and analyzed in more detail in Section 5.4.

Further, in this section it was assumed that all of the workload was processed on workshop machines that were internal to the organization. In many situations advantage can be taken of subcontractors to do some of the work during busy periods. The approach used in Section 5.6 to determine the optimal size of a maintenance crew, taking account of subcontracting opportunities, can, in particular cases, be used to determine the optimal number of workshop machines where subcontracting opportunities occur.

5.3.5 APPLICATIONS

5.3.5.1 Optimizing the Backlog

In a plant there was a crew of plumbers and the goal was to establish the optimal number of plumbers to have to ensure that the backlog of work (jobs in the queue) did not exceed a specified number or, equivalently, did not result in a job waiting greater than a specified average before a plumber was dispatched to attend to the job.

Knowing the average arrival rate of jobs per week, ensuring that the arrivals could be described as arriving according to a Poisson distribution (which was the case in the plant, since plumbing jobs occurred in many areas of the plant), and that the service times could be described by an exponential distribution (which was the case, since most jobs were small ones, with only a few taking a long time to complete a repair), then Wilkinson's queuing chart shown in Figure 5.5 can be used to estimate the average queuing times for different maintenance crew sizes. Figure 5.9 illustrates the problem. A final crew size decision is then set through management specifying an acceptable waiting time in the queue for an incoming plumbing job. The final results of the study are provided in Table 5.2.

5.3.5.2 Crew Size Optimization

There were two maintenance crews with responsibility in a company to handle a task called pulley replacement. The mean arrival rate of pulleys per month to the two teams was 125; the average capacity per month of each team was 102 jobs per month. Hours available for a team to work in a month were 213.

Noting that the arrival rate is 125 and the capacity of one team is 102, clearly at least two teams are required. And with two teams it was estimated using Wilkinson's chart (Figure 5.5) that the average utilization of a team is 61%, and the average waiting time for a pulley replacement request to be attended to was

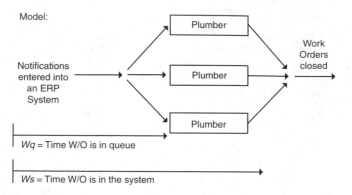

FIGURE 5.9 Backlog optimization.

TABLE 5.2
Service Level Provided by Plumbers

Number of Plumbers, n	Mean Time of a W/O in the Queue, W_q (% of a week)	Mean Time of a W/O in the System, W_s (% of a week)
3	0.080	0.147
4	0.026	0.093
5	0.004	0.071
6	0.002	0.069

1 hour. A decision was made to maintain the two teams, and not explore the addition of a new team.

In this study the cost per month of a maintenance team and the cost of lost production for 1 month were known, so had it been necessary, a formal optimization calculation would have been conducted.

5.4 OPTIMAL MIX OF TWO CLASSES OF SIMILAR EQUIPMENT (SUCH AS MEDIUM/LARGE LATHES) TO MEET A FLUCTUATING WORKLOAD

5.4.1 STATEMENT OF PROBLEM

The problem of this section is an extension of the problem in Section 5.3, which dealt with the optimal number of identical workshop machines to meet a fluctuating demand.

Specifically, in this section we assume that there is a class of machines — lathes used in the workshop — that can be divided into medium and large lathes. Jobs requiring lathe work can then be divided into those that require processing on a medium lathe, those that require a large lathe, or those that can be processed equally well on either. The service times of jobs on medium and large lathes differ, as do the operating costs of the lathes.

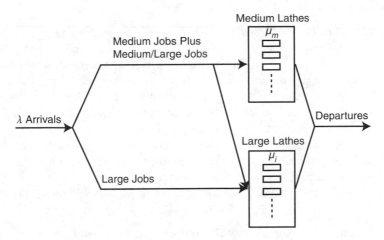

FIGURE 5.10 Workshop system for two classes of equipment.

For a given workload pattern, the problem is to determine the optimal mix of medium/large lathes to minimize the total cost per unit time of the lathes and downtime costs associated with jobs waiting in a queue or being processed.

5.4.2 CONSTRUCTION OF MODEL

Figure 5.10 illustrates the queuing system for the problem. Thus, it is seen that lathe-requiring work arriving at the lathes can be divided into work that requires the use of:

1. A medium lathe (operating cost low)
2. A large lathe (operating cost high)
3. Either a medium or large lathe

To attempt to analyze the above system analytically is not practicable due to the complexity of the mathematics that would be involved. Simulation, however, is a convenient alternative and is readily understandable. We will now introduce this procedure.

Simulation basically consists of four steps:

1. Determine the logic of the system being analyzed and represent it by means of a flowchart.
2. Obtain the information necessary to work through the flowchart.
3. Simulate the operation of the system for different alternatives by using the data obtained in step 2 and working through the logic specified in step 1. The simulation can be done manually or by computer.
4. Evaluate the consequences obtained in step 3, and so identify the best alternative.

5.4.2.1 Logic Flowchart

Since in practice most jobs that can be processed on a medium lathe can also be processed on a large lathe, we consider a two-queue system: one queue at the medium lathes, composed of all jobs requiring at least a medium lathe, and one queue at the large lathes, composed of all jobs requiring only a large lathe.

Whenever a medium lathe becomes vacant, it takes the first job in the medium/large queue and processes it. If there is no queue at the medium lathes, then the medium lathes are idle.

Whenever a large lathe becomes vacant, it takes the first job in the large lathes queue. If there is no queue at the large lathes, then, if possible, a job is transferred from the medium/large queue to the large lathe.

The logic of the system is illustrated in the flowchart of Figure 5.11.

5.4.2.2 Obtaining Necessary Information and Constructing Model

We shall suppose that observations of the system have been made and that the following distributions have been obtained:

1. The arrival of jobs to the lathe system is a Poisson process, with an arrival rate of λ per unit time. Thus, the interarrival distribution of jobs will be negative exponential with a mean interval $1/\lambda$.
2. The probability that an incoming job joins the queue at the medium lathes is p; hence, the probability that the job joins the large lathe queue is $(1-p)$.
3. The service times for jobs on the medium and large lathes are negative exponentially distributed, with mean service rates of μ_m and μ_l per unit time, respectively.
4. The downtime cost per unit time for a job waiting in a queue or being processed is C_d.
5. The cost of operation per unit time for one medium lathe is C_m, and for one large lathe it is C_l.

The objective is to determine the optimal number of medium (n_m) and large (n_l) lathes to minimize the total cost per unit time $C(n_m, n_l)$ associated with the lathes and downtime costs of jobs being in the workshop for repair.

$$C(n_m, n_l) = \text{cost per unit time for medium lathes}$$
$$+ \text{cost per unit time for large lathes}$$
$$+ \text{downtime cost per unit time for jobs waiting}$$
$$\text{or being processed in the medium lathe system}$$
$$+ \text{downtime cost per unit time for jobs waiting}$$
$$\text{or being processed in the large lathe system}$$

$$\text{Cost per unit time for medium lathes} = n_m C_m$$

$$\text{Cost per unit time for large lathes} = n_l C_l$$

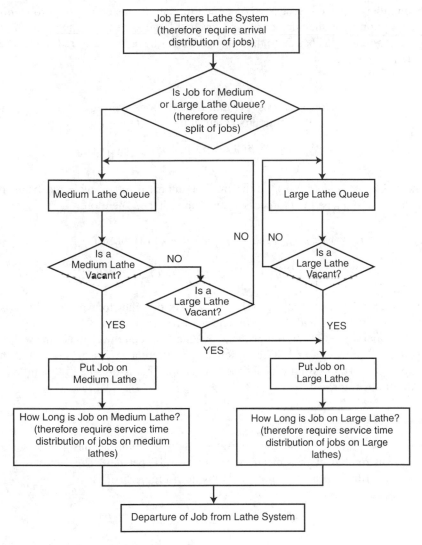

FIGURE 5.11 Flowchart of system structure.

Downtime cost per unit time for jobs waiting
or being processed in medium system
= mean waiting time in system for one job
× arrival rate of jobs to system
× downtime cost per unit time per job

$$= W_{s,m} \times \lambda \times p \times p(n_m, n_l) \times C_d$$

Note: The probability that a job enters the medium system is $p \times p(n_m, n_l)$, where $p(n_m, n_l)$ is the probability that an incoming job that is allocated to the medium/large queue is processed on a medium lathe. This processing probability is dependent upon the number of medium and large lathes. Then the probability that a job initially allocated to the medium/large queue is transferred to the large system is $1 - p(n_m, n_l)$.

Similarly,

$$\text{Downtime cost per unit time for jobs waiting or being processed in large system}$$

$$= W_{s,l}\left\{\lambda \times (1-p) + \lambda \times p \times \left[1 - p(n_m, n_l)\right]\right\}C_d$$

where $\lambda \times p \times \left[1 - p(n_m, n_l)\right]$ is the mean number of jobs transferred from the medium/large queue to be processed on a large lathe. Therefore,

$$C(n_m, n_l) = n_m C_m + n_l C_l + W_{s,m} \times \lambda \times p \times p(n_m, n_l) \times C_d$$
$$+ W_{s,l}\left\{\lambda \times (1-p) + \lambda \times p \times \left[1 - p(n_m, n_l)\right]\right\} \times C_d \quad (5.2)$$

This is a model of the problem relating the mix of lathes to expected total cost. (Note that both $W_{s,m}$ and $W_{s,l}$ are functions of n_m and n_l.)

The major problem in solving the above model is determination of the waiting times in the medium and large systems, for different mixes of lathes and the corresponding processing probabilities $p(n_m, n_l)$. This is obtained by simulation in the following example.

5.4.3 Numerical Example

1. The number of jobs arriving at the lathe section per day is Poisson distributed, with a mean arrival rate of 10 per day. The cumulative distribution function for this is given in Figure 5.12.

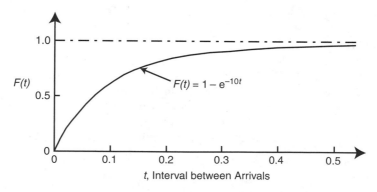

FIGURE 5.12 Cumulative distribution function: job arrivals.

2. The probability that an incoming job joins the queue at the medium lathes is 0.8.
3. The service distribution for jobs on a medium lathe is negative exponential, with a mean service rate for the lathe of 2 per day. The cumulative distribution function for this is given in Figure 5.13.
4. The service time distribution for jobs on a large lathe is negative exponential, with a mean service rate for the lathe of 1 per day. The cumulative distribution function for this is given in Figure 5.14.
5. The downtime cost per job C_d is $1 per day.
6. The costs of operation C_m and C_l are $7 and $10 per day, respectively.
7. The queuing times for jobs at the medium and large lathes are obtained by simulation as follows.

First, we must assume a certain number of medium and large lathes. We might estimate this as follows: There are 10 jobs per day, on average, arriving at the lathes. Eighty percent require processing on a medium lathe. Therefore, an average of eight jobs per day require a

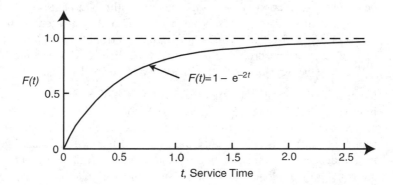

FIGURE 5.13 Cumulative distribution function: service times on medium lathes.

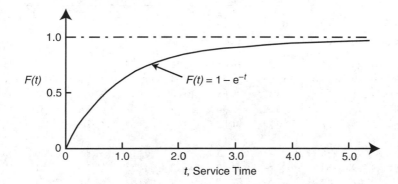

FIGURE 5.14 Cumulative distribution function: service times on large lathes.

medium lathe, and two jobs per day require a large lathe. A medium lathe can process, on average, two jobs per day. A large lathe can process, on average, one job per day.

Let us assume that we have four medium lathes and three large lathes. Note: If we only had two large lathes, which might appear to be sufficient, then the traffic intensity of the system ρ would be 1. As we have seen (sample calculation of Section 5.3.3), this would lead to infinite waiting times.

With reference to the logic flowchart (Figure 5.11):

1. Assume job 1 arrives at the lathe at time 0.
2. Select a number between 00 and 99 from a table of random sampling numbers (see Table 5.3 for an extract). If it is <80, the job goes to the medium queue; otherwise, it goes to the large queue. Taking the first two-digit number in Table 5.3, we get 20; therefore, job 1 goes to the medium lathes.
3. Select another number from Table 5.3. This number is now used to determine the duration of job 1 on a medium lathe. The next two-digit number in row 1 is 17. This is taken as 0.17 and is marked on the y-axis of Figure 5.13. Drawing a horizontal line from 0.17 until it cuts the $F(t)$ curve, then dropping a vertical line, gives a service time of 0.10 day as being equivalent to a random number of 17.

Note: In this example the random sampling numbers are taken to be in the range 0.005 to 0.995 in steps of 0.01 to preclude the possibility of a zero or infinite service time being specified. The extract of random sampling numbers is taken from Lindley and Miller (1964). Each digit is an independent sample from a population in which the digits 0 to 9 are equally likely; that is, each has a probability of 1/10.

Thus, a random number of 17 is equivalent to $F(t) = 0.175$. So $0.175 = 1 - e^{-2t}$, and therefore $t = 0.10$ day.

TABLE 5.3
Random Numbers

20 17	42 28	23 17	59 66	38 61	02 10	86 10	51 55	92 52	44 25
74 49	04 49	03 04	10 33	53 70	11 54	48 63	94 60	94 49	57 38
94 70	49 31	38 67	23 42	29 65	40 88	78 71	37 18	48 64	06 57
22 15	78 15	69 84	32 52	32 54	15 12	54 02	01 37	38 37	12 93
93 29	12 18	27 30	30 55	91 87	50 57	58 51	49 26	12 53	96 40
45 04	77 97	36 14	99 45	52 95	69 85	03 83	51 87	85 56	22 37
44 91	99 49	89 39	94 60	48 49	06 77	64 72	59 26	08 51	25 57
16 23	91 02	19 96	47 59	89 65	27 84	30 92	63 37	26 24	23 66
04 50	65 04	65 65	82 42	70 51	55 04	61 47	88 83	99 34	82 37
32 70	17 72	03 61	66 26	24 71	22 77	88 33	17 78	08 92	73 49

4. As there are no other jobs in the system, we can put job 1 straight onto a medium lathe, say m_1, the first medium lathe, for 0.1 day.

   All the above information is given in the first row of Table 5.4.
5. We now have to generate the arrival of another job. To do this, we select a further random number, in this case 42 from the top row of Table 5.3. Marking 0.42 on the y-axis of Figure 5.12, we get an equivalent interval between job 1 and job 2 of 0.06 day from the x-axis.

Proceeding as indicated in steps 2 to 4 above, the second row of Table 5.4 can be completed. The interval between arrivals of job 2 and job 3 can be obtained as indicated in step 5 above, and row 3 of Table 5.4 can be completed as per steps 2 to 4 above. Similarly, rows 4 to 7 of the table can be completed.

Clearly, the construction of a table such as Table 5.4 by hand is tedious. However, if we proceeded as above, we would eventually generate sufficient jobs to obtain the average waiting time (from columns 6 and 7) for jobs in the medium or large lathe systems when there are four medium and three large lathes and the probability $p(4, 3)$ of jobs being processed on the medium lathes (from columns 4 and 5). To reduce the tedium and speed up the calculations, it is usually possible to take advantage of one of the many simulation packages that are available to perform the simulation, such as Arena, Witness, SimProcess, and Micro Saint. A common feature of these simulation packages is the ability to use animation to show the dynamic behavior of the modeled system during a simulation run. With this capability, the number of jobs waiting to be processed by medium lathes, for example, can be shown graphically instead of indicated by a number. This helps the user to validate the model and to understand the behavior of the model more easily.

The simulation model presented in Section 5.4.2 is implemented in the Workshop Simulator software that can be downloaded from http://www.crc press.com/e_products/downloads/download.asp?cat_no=DK9669. The results of a simulation run after entering the data of this numerical example into the software are shown in Figure 5.15.

Using the workshop simulator software or one of the other simulation packages introduced above, the average results of several simulations can be obtained as listed in Table 5.5, which gives the mean waiting time results for the data used in construction of Table 5.4 and the processing probability $p(4, 3)$.

Table 5.6 gives the appropriate mean waiting times and processing probabilities for other feasible combinations of numbers of medium and large lathes, that is, ones that result in a steady state.

Once the waiting times and probabilities have been determined, solutions to the model can be obtained (Equation 5.2). Table 5.7 gives the various total costs per day, and it is seen that the optimal mix is five medium and three large lathes.

Sample calculation:

When $n_m = 4$, $n_l = 3$, $p = 0.8$, and $\lambda = 10$, from the simulation we obtain $W_{s,m}$ = 0.79 days and $W_{s,l}$ = 1.08 days. These are the mean times that jobs processed on medium and large lathes spent in the system.

TABLE 5.4
Manual Simulation Results

1 Job No.	2 Interarrival Time between Jobs	3 Cumulative Time	4 Queue Decision	5 Is a Suitable Lathe Vacant?	6 Waiting Time in Queue	7 Service Time on Lathe (days)	8 Lathe Used	9 Starting Time on Lathe	10 Finishing Time on Lathe (Cumulative Time)	11 Next Job on Lathe
1		0.00	r.n. = 20 — m queue	Yes	0	r.n. = 170.10	m_1	0.00	0.10	6
2	r.n. = 420.06	0.06	r.n. = 28 — m queue	Yes	0	r.n. = 230.13	m_2	0.06	0.06 + 0.12 = 0.19	7
3	r.n. = 170.02	0.08	r.n. = 59 — m queue	Yes	0	r.n. = 660.55	m_3	0.08	0.08 + 0.55 = 0.63	
4	r.n. = 380.05	0.13	r.n. = 61 — m queue	Yes	0	r.n. = 20.01	m_4	0.13	0.13 + 0.01 = 0.14	
5	r.n. = 100.01	0.14	r.n. = 86 — m queue	Yes	0	r.n. = 110.11	l_1	0.14	0.14 + 0.11 = 0.25	
6	r.n. = 510.07	0.21	r.n. = 55 — m queue	Yes (m_1 is vacant at time 0.10)	0	r.n. = 921.30	m_1	0.21	0.21 + 1.30 = 1.51	
7	r.n. = 520.07	0.28	r.n. = 44 — m queue	Yes (m_2 is vacant at time 0.19)	0	r.n. = 250.15	m_2	0.28	0.28 + 0.15 = 0.43	

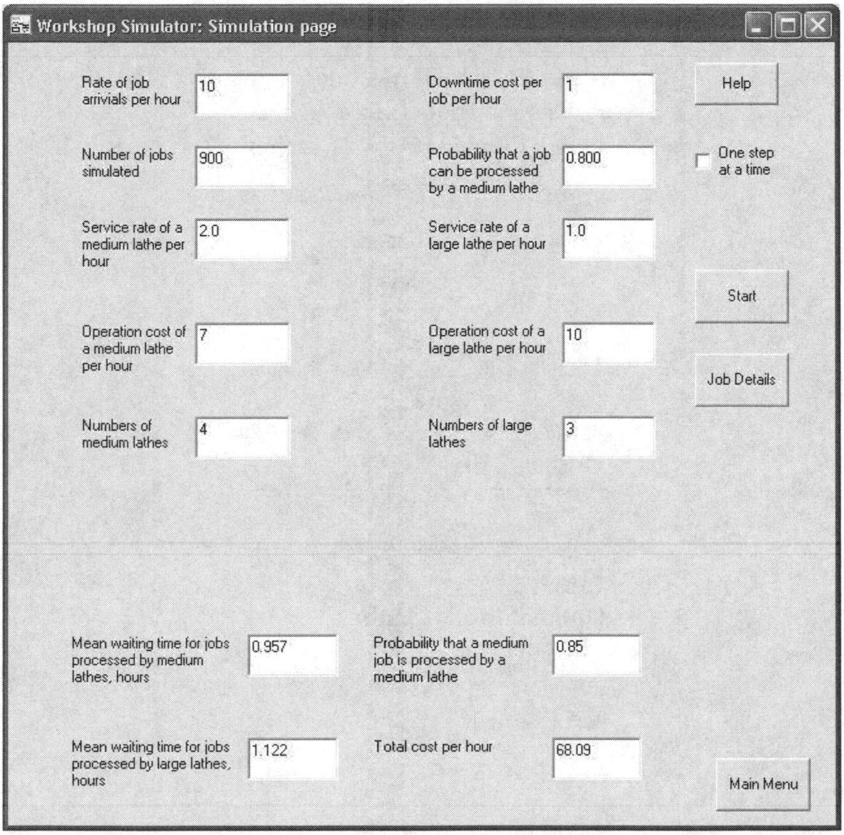

FIGURE 5.15 Input and output of a simulation run of the Workshop Simulator software.

TABLE 5.5
Mean Waiting Times and
Processing Probability, $p(4, 3)$

$n_m = 4$ $W_{s,m} = 0.79$ $p(4, 3) = 0.82$
$n_l = 3$ $W_{s,l} = 1.08$

The probability that a job that is allocated to the medium/large queue on entry to the system is processed on a medium lathe, $p(4, 3)$, is obtained as 0.82. Therefore, the probability that a job is switched from the medium/large queue to be processed on a large lathe is $1 - 0.82 = 0.18$. We therefore obtain

$$C(4,3) = 4 \times 7 + 3 \times 10 + 0.79 \times (10 \times 0.8 \times 0.82) \times 1 + 1.08 \times$$
$$(10 \times 0.2 + 10 \times 0.8 \times 0.18) \times 1 = \$66.90 \text{ per day}$$

TABLE 5.6
Mean Waiting Times and
Processing Probabilities for
Other Combinations of Lathes

$n_m = 5$	$W_{s,m} = 0.58$	$p(5, 3) = 0.89$
$n_l = 3$	$W_{s,l} = 0.98$	
$n_m = 6$	$W_{s,m} = 0.54$	$p(6, 3) = 0.94$
$n_l = 3$	$W_{s,l} = 1.08$	
$n_m = 4$	$W_{s,m} = 0.48$	$p(4, 4) = 0.82$
$n_l = 4$	$W_{s,l} = 0.87$	
$n_m = 5$	$W_{s,m} = 0.55$	$p(5, 4) = 0.85$
$n_l = 4$	$W_{s,l} = 0.83$	
$n_m = 6$	$W_{s,m} = 0.52$	$p(6, 4) = 0.90$
$n_l = 4$	$W_{s,l} = 0.87$	

TABLE 5.7
Optimal Mix of Lathes

(n_m, n_l)	$C(n_m, n_l)$
$n_m = 4, n_l = 3$	66.88
$n_m = 5, n_l = 3$	71.97
$n_m = 6, n_l = 3$	78.74
$n_m = 4, n_l = 4$	75.04
$n_m = 5, n_l = 4$	81.40
$n_m = 6, n_l = 4$	88.15

Note: For each combination of medium and large lathes, four simulation runs were made on the computer to obtain the waiting time statistics. Each run was equivalent to 2 months operation.

5.4.4 FURTHER COMMENTS

Simulation is a very useful procedure for tackling complex (and not so complex) queuing type problems. The reader interested in a fairly complete discussion of the subject is referred to Law and Kelton (2001) or Banks et al. (1995).

In the model it was assumed that the processing time for a job that could be done on a medium lathe, but that was switched to a large lathe, could be taken from the same service time distribution as a job requiring processing on a large lathe. This may be realistic, since medium jobs may require longer setting up times on a large lathe, and thus offset the increased speed of doing the job on the larger lathe. However, if this assumption is not acceptable, then the model would need to be modified. Also, in the model it was implied that the operating cost of a lathe was constant and independent of whether the lathe was being used. Removal of these assumptions is not difficult, but a more complicated model would result than the one discussed in this section.

It will be appreciated that in construction of Table 5.7, the appropriate mixes of medium and large lathes to use in the simulation were obtained on a subjective basis — through careful thought about the consequences resulting from previously tried combinations. Thus, it is obvious that the use of simulation may result in the optimum being missed, since it is often not feasible to attempt to evaluate all possible alternatives. In practice, this is not usually a severe restriction.

Another problem with simulation is deciding just how long a simulation run should last and how many runs should be made, since it is only after a sufficiently large number of sufficiently long runs are made that the steady state is reached and averages can be calculated and used in a model. Discussion of the cutoff point, and other aspects of experimental design, is covered in the textbooks referred to at the beginning of this section.

For simple problems, a hand simulation may be worthwhile, and if this is done, tables of random numbers will be required. Table 5.3 is an extract of such tables that appear in many books of statistical tables. Tables of random numbers consist of a sequence of the digits 0, 1, …, 9, having the property that any position in the sequence has an equal probability of containing any one of these 10 digits. Such sequences can be broken down into subsequences of n digits having the same property. Suppose it is necessary to draw an item at random from a population of 1000 items. If these items are imagined to carry labels with numbers ranging from 000 to 999, selecting an item at random is then equivalent to selecting a three-digit number at random. This condition is satisfied by entering the table at any point and selecting the item corresponding to that number in the table. Repetition of this process allows a random sample of any desired size to be selected provided that the three-digit numbers taken from the table are accepted every time in the same sequence. It should be noted that the same functionality as tables of random numbers is provided by the RND function on most calculators. Each time it is pressed, it generates a different pseudorandom number uniformly distributed over the range 0 to 1.

Conversion of random sampling numbers to random variables (as is done in the simulation example of Section 5.4.3) is done via the appropriate cumulative distribution function of the variable. Once a random sampling number is obtained (from tables), the corresponding value of the random variable is read off the distribution function (see Section 5.4.3). In the example, two-digit random sampling numbers were used (in the range of 00 to 99) and then taken to be in the range of 0.005 to 0.995 in steps of 0.01.

Although the example of this section dealt specifically with determination of the optimal mix of lathes in a workshop, the approach is applicable to other maintenance problems. For example, a problem that frequently occurs is the necessity of determining the appropriate skills to have available in a maintenance team and the number of craftspeople possessing these skills. Certain jobs can be tackled equally well by any member of the team, while others require specialists. The different classes of skills that can be defined will almost certainly exceed two, but even so, the optimal mix of these skills can be determined in a manner similar to that of this section.

Finally, in the model it has been assumed that downtime cost could be obtained. As is often the case, this is a difficult costing problem, and so the analysis of such a problem may stop at determining the consequences in terms of waiting times for different mixes of lathes, with management then deciding which alternative it prefers on the basis of the calculated waiting times.

5.4.5 APPLICATIONS

5.4.5.1 Establishing the Optimal Number of Lathes in a Steel Mill

Within an integrated steel mill there was the need to deliver an improved service level to operations. The present practice was to plan to process small jobs on small lathes and large jobs on large lathes. However, if a large lathe became vacant, and there was a queue of jobs waiting for processing on a small lathe, the workshop planner would transfer a small job and have it processed on a large lathe. It was not feasible to transfer a large job and process it on a small lathe. Only transfers upward were possible.

The lathe section consisted of eight lathes, and Figure 5.16 shows the initial perspective of the lathe system: it consisted of two classes of lathes, one termed large and the other medium.

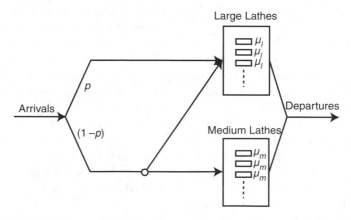

FIGURE 5.16 Initial perspective of lathe system.

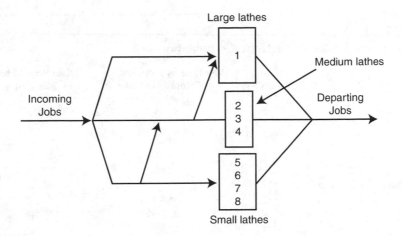

FIGURE 5.17 Second view of lathe system.

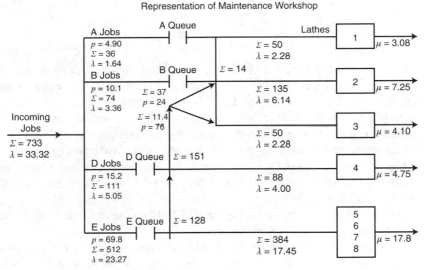

p = Percentage of jobs going along branch; Σ = Total number of jobs going along branch in 22 weeks;
λ = Arrival rate of jobs / week during 22 week period; μ = Service rate of jobs / week during 22 week period

FIGURE 5.18 Final view of lathe system.

As data were acquired, it became clear that the system was more complex than initially defined, and the improved understanding of the lathe system is depicted in Figure 5.17.

After further data collection and analysis over a 22-week period, it was clear that the system could be represented by Figure 5.18, in which various statistics are provided.

TABLE 5.8
Evaluation of Alternative Lathe Combinations

Situation	Lathe Number	Idle Time as a Percentage of Hours Available	Mean Wait of Jobs in Queue (days)
Existing state	1	26.4%	6.55
	2	15.4%	5.25
	3	15.5%	9.37
	4	15.7%	8.10
	5–8	2.0%	5.11
Additional small machine	1	44.6%	2.95
	2	38.8%	1.65
	3	43.0%	1.97
	4	36.4%	2.45
	5–8 +1	2.0%	2.83
Additional type 4 machine	1	44.6%	2.95
	2	42.2%	1.49
	3	39.4%	2.56
	4 + 1	12.6%	2.46
	5–8	20.8%	1.77

It will be noted in Figure 5.18 that the simulation allowed the possibility of a small job (one that could be processed on lathe 5, 6, 7, or 8) being processed on the largest lathe, lathe 1. In practice, this may not be feasible, and if not, the simulation model needs to block such a transfer. In the study described that was not necessary, and when subsequently examining the simulation statistics, it was noted that no simulation run had a small lathe job being processed on the largest lathe.

Once a simulation model is built, it has to be tested and verified (for details, see Law and Kelton [2001] or Banks et al. [1995]) before alternative system configurations can be evaluated.

Once the system illustrated by Figure 5.18 had been tested and verified, the final results given in Table 5.8 were obtained.

Insights obtained by the performance statistics given in Table 5.8 — columns 3 and 4 — can then be used to assist management in deciding on the best combination of lathes to have in its operation. Note again that no formal optimization has taken place. There is simply an explanation of the expected consequences associated with alternative configurations, and management will then consider the costs of the alternatives, and possibly the way the work will be distributed among the different lathe groupings, before a final decision is made.

5.4.5.2 Balancing Maintenance Cost and Reliability in a Thermal Generating Station

A thermal generating station was 25 years old and a decision was made to spend a substantial amount of money on replacement equipment to improve the reliability

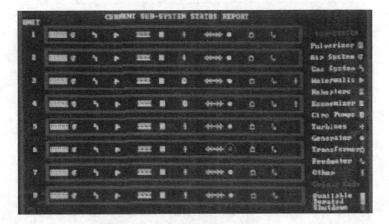

FIGURE 5.19 Generating station simulation schematic.

of the station. The station was one that was brought into service to meet peak electricity demands, but because of the age of its equipment, it occasionally could not meet the demand placed on it.

Within the electrical utility there were a number of alternative suggestions on how to best spend the allocated $300 million, such as replace pulverizers or replace transformers. A decision was made to build a simulation model of the plant and evaluate the alternative suggestions. Subsequently, a particular course of action was implemented. Figure 5.19 is a representation of the generating station used in the simulation, from which it can be seen that there are eight generating units.

Before evaluating alternatives, it is necessary to validate the simulation model, and that is done by comparing historical key performance indicators (KPIs) with results obtained from the simulation. Once an acceptably similar set of values is obtained, the simulation of alternative system designs can commence. New operating cost estimates and equipment failure statistics are used for equipment that is planned to be replaced. Much sensitivity checking is undertaken, with the goal being not so much to get absolute values of the expected future value of KPIs, but to discriminate among the alternatives being evaluated. Table 5.9 illustrates validation of the simulation model.

Further details are provided in Concannon et al. (1990).

5.5 RIGHTSIZING A FLEET OF EQUIPMENT: AN APPLICATION

A given workload had to be completed by a fleet of owner-operated equipment. The work can be undertaken by a small fleet that is highly utilized and whose operation and maintenance cost will be high, or a larger fleet that is not so highly utilized and whose operations and maintenance (O&M) costs will be lower, and

TABLE 5.9
Simulation Model Validation

	Actual Results	Simulation Results
Energy produced (MW)	3763	4098
Total operating (hours)	22,089	19,623
Equivalent F.O. (hours)	7235	5064
CAWN (%)	87.20	91.80
DAFOR	25.70	23.70
F.O. occurrences	351	298

Quite similar

Note: Results compared to actual data for previous year. F.O. = forced outages; CAWN = capacity available when needed; DAFOR = derating adjusted forced outage rate.

thus the economic life would be greater than the highly utilized equipment. What is the best alternative?

Since the solution to this resource requirement problem will draw on the economic life model of Section 4.2.4 (Equation 4.2), "Construction of Model" and "Numerical Example" sections are not included. Only the application is presented.

5.5.1 An Application: Fleet Size in an Open-Pit Mine

The demand in an open-pit mining operation was to use a fleet of a specified size of haul truck to deliver 108,000 hours of work to a mill to provide the required tonnage per year. There are 8760 hours in a year, and so the workload was judged to be able to be undertaken by:

Alternative 1: 16 trucks each delivering 7000 hours per year
Alternative 2: 18 trucks each delivering 6000 hours per year
Alternative 3: 22 trucks each delivering 5000 hours per year

The trend in O&M costs for the trucks is given in Figure 5.20. Knowing this trend, the O&M costs for a truck undertaking H hours per year is

$$\int_{i}^{i+H} c(k)\,dk, \ i = 0, H, 2H, 3H, \text{ etc.}$$

The estimated resale values were also obtained for trucks undertaking 7000, 6000, and 5000 hours per year. Along with the interest rate appropriate for discounting, the economic life and associated equivalent annual costs (EACs) were then obtained for alternatives 1 to 3. These are provided in Table 5.10. While

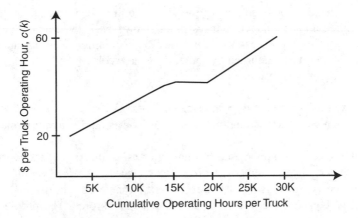

FIGURE 5.20 Trend in truck O&M costs.

TABLE 5.10
Fleet Size Optimization

Fleet Size	Economic Life (years)	EAC ($)	Total Annual Fleet Cost ($)
16	5	506,102	8,097,632
18	5	444,174	7,995,132 (min)
22	6	381,299	8,388,578

the EAC is smallest for an annual utilization of 5000 hours, if this is used as the basis of establishing fleet size, then 22 trucks would be required. Thus, in this class of problem it is clear that what is important is not the economic life of an individual truck, but the fleet as a whole that is required to be optimized. (The same issue arose in Chapter 4 when establishing the economic life of a given fleet whose utilization varied as it aged — Section 4.4.) Examination of Table 5.10 indicated that the optimal fleet size is 18, with each truck delivering 6000 hours of work per year.

5.6 OPTIMAL SIZE OF A MAINTENANCE WORKFORCE TO MEET A FLUCTUATING WORKLOAD, TAKING ACCOUNT OF SUBCONTRACTING OPPORTUNITIES

5.6.1 STATEMENT OF PROBLEM

The workload for the maintenance crew is specified at the beginning of a period, say a week. By the end of the week all the workload must be completed. The size of the workforce is fixed; thus, there is a fixed number of staff available per week. If demand at the beginning of the week requires fewer staff than the fixed capacity, then no subcontracting takes place. However, if the demand is greater than the capacity, the excess workload will be subcontracted to an alternative service deliverer, to be returned by the end of the week.

Two sorts of costs are incurred:

1. Fixed cost depending on the size of the workforce
2. Variable cost depending on the mix of internal and external workload

As the fixed cost is increased through increasing the size of the workforce, there is less chance of subcontracting being necessary. However, there may frequently be occasions when fixed costs will be incurred, yet demand may be low, that is, considerable underutilization of the workforce. The problem is to determine the optimal size of the workforce to meet a fluctuating demand to minimize expected total cost per unit time.

5.6.2 Construction of Model

1. The demand per unit time is distributed according to a probability density function $f(r)$, where r is the number of jobs.
2. The average number of jobs processed per person per unit time is m.
3. The total capacity of the workforce per unit time is mn, where n is the maintenance crew size.
4. The average cost of processing one job by the workforce is C_w.
5. The average cost of processing one job by the subcontractor is C_s.
6. The fixed cost per crew member per unit time is C_f.

The basic conflicts of this problem are illustrated in Figure 5.21, from which it is seen that the expected total cost per unit time $C(n)$ is

$C(n)$ = fixed cost per unit time
+ variable internal processing cost per unit time
+ variable subcontracting processing cost per unit time

Fixed cost per unit time = size of workforce × fixed cost per crew member = nC_f

$$\text{Variable internal processing cost per unit time} = \text{average number of jobs processed internally per unit time} \times \text{cost per job}$$

Now, the number of jobs processed internally per unit time will be:

1. Equal to the capacity when demand is greater than capacity
2. Equal to the demand when demand is less than or equal to capacity

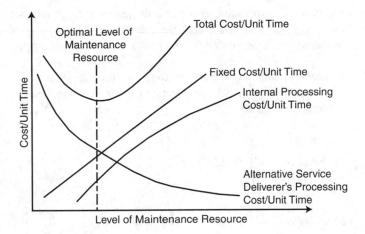

FIGURE 5.21 Optimal contracting-out decision.

The probability of 1 $= \int_{nm}^{\infty} f(r)dr$.

The probability of 2 $= \int_{0}^{nm} f(r)dr = 1 - \int_{nm}^{\infty} f(r)dr$.

When 2 occurs, the average demand will be

$$\frac{\int_{0}^{nm} rf(r)dr}{\int_{0}^{nm} f(r)dr}$$

Therefore, the variable internal processing cost per unit time

$$= \left(nm \int_{nm}^{\infty} f(r)dr + \frac{\int_{0}^{nm} rf(r)dr}{\int_{0}^{nm} f(r)dr} \times \int_{0}^{nm} f(r)dr \right) C_w$$

Variable subcontracting processing cost per unit time = average number of jobs processed externally per unit time × cost per job

Now, the number of jobs processed externally will be:

1. Zero when the demand is less than the workforce capacity
2. Equal to the difference between demand and capacity when demand is greater than capacity

The probability of 1 $= \int_{0}^{nm} f(r)dr$.

The probability of 2 $= \int_{nm}^{\infty} f(r)dr = 1 - \int_{0}^{nm} f(r)dr$.

When 2 occurs, the average number of jobs subcontracted is

$$\frac{\int_{nm}^{\infty}(r-nm)f(r)dr}{\int_{nm}^{\infty}f(r)dr}$$

In this case, the variable subcontracting processing cost per unit time

$$= \left(0 \times \int_{0}^{nm} f(r)dr + \frac{\int_{nm}^{\infty}(r-nm)f(r)dr}{\int_{nm}^{\infty}f(r)dr} \times \int_{nm}^{\infty} f(r)dr \right) C_s$$

Therefore,

$$C(n) = nC_f + \left(nm\int_{nm}^{\infty} f(r)dr + \int_{0}^{nm} rf(r)dr \right)C_w + \left(\int_{nm}^{\infty}(r-nm)f(r)dr \right)C_s \quad (5.3)$$

This is a model of the problem relating workforce size n to total cost per unit time $C(n)$.

5.6.3 NUMERICAL EXAMPLE

1. It is assumed that the demand distribution of jobs per week can be represented by a rectangular distribution having the range of 30 to 70; i.e., $f(r) = 1/40$, $30 \leq r \leq 70$, $f(r) = 0$ elsewhere.
2. $m = 10$ jobs per week, $C_w = \$2$, $C_s = \$10$, $C_f = \$40$.

Equation 5.3 becomes

$$C(n) = \$40n + \left(10n \int_{10n}^{70} \frac{1}{40} \, dr + \int_{30}^{10n} \frac{r}{40} \, dr \right) \times \$2 + \left(\int_{10n}^{70} \frac{r - 10n}{40} \, dr \right) \times \$10$$

Table 5.11, which gives the values of $C(n)$ for all possible values of n, indicates that for the costs used in the example, the optimal solution is to have a maintenance crew of five.

TABLE 5.11
Optimal Crew Size

n	0	1	2	3	4	5	6	7
$C(n)$	500	460	420	380	350	340	350	380

Sample calculation:

When $n = 5$,

$$C(5) = \$200 + \left(50 \times \int_{50}^{70} \frac{1}{40} \, dr + \int_{30}^{50} \frac{r}{40} \, dr \right) \times \$2 + \left(\int_{50}^{70} \frac{r - 50}{40} \, dr \right) \times \$10$$

$$= \$200 + (50 \times 0.5 + 20) \times \$2 + 5 \times \$10 = \$340$$

The decision model presented in Section 5.6.2 is implemented in the Crew Size Optimizer software that can be downloaded from http://www.crc press.com/e_products/downloads/download.asp?cat_no=DK9669. Figure 5.22 shows the results of analysis produced by this software given the data of this numerical example.

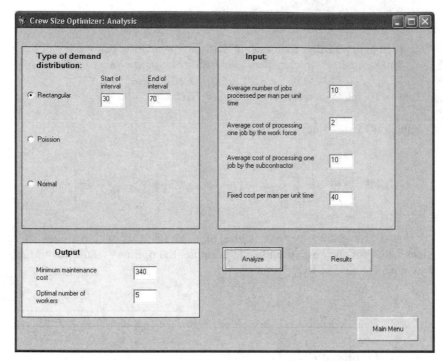

FIGURE 5.22 Input and output screen of the Crew Size Optimizer software.

Pressing the "Analyze" button on the "Input and Output" screen of the software will produce what is shown in Figure 5.23, which presents in graphical and tabular forms the costs for different sizes of the workforce.

5.6.4 FURTHER COMMENTS

In construction of the model of this section, it was assumed that all jobs requiring attention at the start of the week had to be completed by the end of the week. In practice, this requirement would not be necessary if jobs could be carried over from one week to another, that is, backlogged. Inclusion of this condition would result in a model more complicated than that of Equation 5.3.

5.6.5 AN EXAMPLE: NUMBER OF VEHICLES TO HAVE IN A FLEET (SUCH AS A COURIER FLEET)

If the demand per day of deliveries by the company is described by Figure 5.24, and all deliveries have to be completed the same day, we can define:

f = fixed cost to the company for one vehicle per day
h = rental cost of one vehicle per day

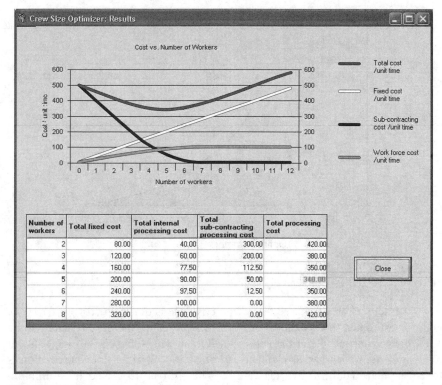

FIGURE 5.23 Crew Size Optimizer: results of analysis.

v = variable cost per day for using one owned vehicle

N = number of vehicles owned by the company

Then it can be derived from Equation 5.3 that the optimal number of vehicles to own, N^*, must satisfy the following inequality:

$$\int_{0}^{N^*} f(n)\,dn \geq \frac{f}{(h-v)} \tag{5.4}$$

This is illustrated graphically in Figure 5.24, where the right-hand side of Equation 5.4 is given by the shaded area.

5.7 THE LEASE OR BUY DECISION

5.7.1 STATEMENT OF PROBLEM

The following example is taken from Theusen (1992), where the following point is made: "The economic advantage of owning or leasing can be determined by evaluating the after-tax cash flow that is associated with each of the options." The problem presented by Theusen follows.

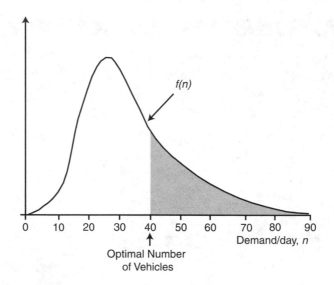

FIGURE 5.24 Delivery demand distribution.

A vehicle when purchased has a first cost (the acquisition cost) of $10,000, and it is estimated that its annual operating expenses for the next 4 years will be $3000 per year, payable at the end of each year. The vehicle's salvage value at the end of the fourth year is estimated to be $2000. Straight-line depreciation is to be applied. The effective tax rate for the firm is 45%, and the minimum attractive rate of return (MARR) is 15%. (This is the same as the inflation-free interest rate discussed in Appendix 3.)

The following three alternatives are to be evaluated:

1. Purchase the vehicle, using retained earnings.
2. Purchase the vehicle, using borrowed funds.
3. Lease the vehicle.

Which is the best alternative?

5.7.2 SOLUTION OF PROBLEM

5.7.2.1 Use of Retained Earnings

The cash flows, before tax, associated with this alternative are given in Figure 5.25.

Since $2000/year depreciation is allowed, the effective cost in years 1 to 4 is $3000 for operating, plus $2000 for depreciation, giving a total of $5000. But since we can write off this cost against taxes, our tax savings amounts to $5000 × (0.45) = $2250, and the next cost is $3000 − $2250 = $750 per year. The after-tax cash flow picture is shown in Figure 5.26.

Using the standard discounting approach as described in Appendix 3 and used in Chapter 3, when the interest rate to be used for discounting is 15%, the EAC associated with the cash flows of Figure 5.26 is $3852.

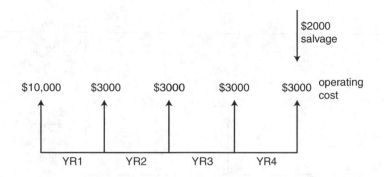

FIGURE 5.25 Use of retained earnings: before-tax cash flows.

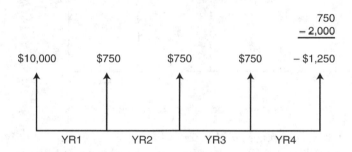

FIGURE 5.26 Use of retained earnings: after-tax cash flows.

5.7.2.2 Use of Borrowed Funds

When funds are borrowed, the interest to be repaid by the borrower can be deducted as an expense before taxes are calculated.

Assuming $10,000 is borrowed at a cost of 18%, and that for the first 3 years of the loan interest only will be repaid, with the principal and last year's interest paid at the end of the fourth year, the cash flows (before tax) are as depicted in Figure 5.27.

Since tax is at 45%, the tax savings amounts to 45% of $4800 = $2160.

A further tax saving is due because of the $2000 depreciation allowance ($2000 × 0.45 = $900). Thus, the total tax benefit = $3060.

The after-tax cash flow picture is shown in Figure 5.28. Note: $4800 − $3060 = $1740 and $10,000 + $4800 − $2000 − $3060 = $9740.

Again, using the standard discounting approach as described in Appendix 3 and used in Chapter 3, when the interest rate to be used for discounting is 15%, the EAC associated with the cash flows of Figure 5.28 is $3342.

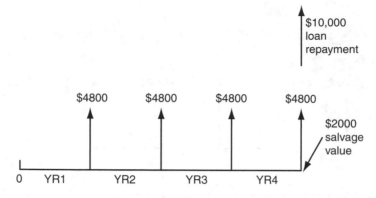

FIGURE 5.27 Use of borrowed funds: before-tax cash flows.

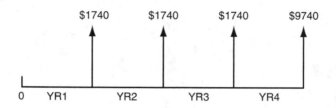

FIGURE 5.28 Use of borrowed funds: after-tax cash flows.

In fact, borrowed funds are cheaper as long as the after-tax cost of borrowing $[(1 - 0.45) \times 18\% = 9.9\%]$ remains less than the after-tax MARR (i.e., 15%).

5.7.2.3 Leasing

If the vehicle is leased by the borrower, no depreciation for tax purposes is allowed. Assuming that the lease requires yearly payments of $2900, with the lessee also paying the $3000 operating cost, the cash flows before tax considerations are as given in Figure 5.29.

Taking tax benefits into account of $0.45 \times \$5900 = \$2655$, then the after-tax cash flows are given in Figure 5.30. Note, $5900 – $2655 = $3245.

Again, discounting the cash flows of Figure 5.29 using $i = 15\%$, the EAC is obtained as $3245.

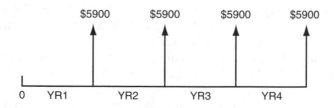

FIGURE 5.29 Leasing: before-tax cash flows.

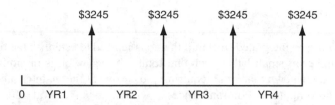

FIGURE 5.30 Leasing: after-tax cash flows.

5.7.2.4 Conclusion

The best alternative is to lease, since $3254 is lower than the cost associated with purchasing the vehicle, using either retained earnings or borrowed funds.

5.7.3 FURTHER COMMENTS

It must be carefully noted that the conclusion for this example should not be generalized. Each individual lease/buy decision should be carefully evaluated using data relevant to the particular situation being analyzed. Also, great care needs to be taken when evaluating alternatives on an after-tax basis to ensure that all the appropriate tax rulings are applied. Lending organizations often have customized software available to undertake the lease/buy evaluation.

REFERENCES

Banks, J., Carson, J.S., and Nelson, B.L. (1995). *Discrete-Event System Simulation*, 4th ed. Englewood Cliffs, NJ: Prentice Hall.

Carruthers, A.J., MacGow, I., and Hackemer, G.C. (1970). A study on the optimum size of plant maintenance gangs. In *Operational Research in Maintenance*, A.K.S. Jardine, Ed. Manchester, U.K., Manchester University Press.

Concannon, K.H., Jardine, A.K.S., and McPhee, J. (1990). Balance maintenance cost against plant reliability with simulation modeling. *Industrial Engineering*, January, pp. 24–27.

Cox, D.R. and Smith, W.L. (1961). *Queues*. London: Chapman & Hall.

Gross, D. and Harris, C. (1997). *Fundamentals of Queueing Theory*, 3rd ed. New York: Wiley.

Law, A.M. and Kelton, W.D. (2001). *Simulation Modeling and Analysis*, 3rd ed. New York: McGraw-Hill.

Lindley, D.V. and Miller, J.C.P. (1964). *Cambridge Elementary Statistical Tables*. Cambridge, U.K., Cambridge University Press.

Morse, P.M. (1963). *Queues, Inventories and Maintenance*. New York: Wiley.

Peck, C.G. and Hazelwood, R.N. (1958). *Finite Queueing Tables*. New York: Wiley.

Theusen, G.J. (1992). Project selection and analysis. In *Handbook of Industrial Engineering*, G. Salvendy, Ed. New York: Wiley, chap. 51.

Wilkinson, R.I. (1953). Working curves for delayed exponential calls served in random order. *Bell System Technical Journal*, 32.

PROBLEMS

1. Within an integrated steel mill there is the need to establish the optimal number of small lathes such that total cost per week is minimized.

    It is known that lathe-requiring jobs arrive at the maintenance shop according to a Poisson process, with a mean arrival rate of 25 jobs/week. Taking into account the productivity characteristics of the lathes, it is known that an appropriate estimate of the mean time to process one job is exponentially distributed with a mean time of 1/7 week.

    Work tied up in the workshop while waiting for processing or being processed costs $10,000 per week, and the total cost per week for one lathe, including operator and overhead, is $5000.

    Establish the optimal number of lathes such that the total cost associated with the lathes and duration associated with the turnaround time of jobs in the workshop is minimized.

2. Suppose you have to decide on the number of crew members to hire for the provisioning of maintenance services. The wage of each crew member is C_w per week, and the cost of completing one maintenance job by the in-house crew is C_r. In cases where the service demand exceeds in-house capacity, the unmet demand will be contracted out. Two contracting options are available: Option 1, at a fixed unit cost of $C_1 > C_r$; Option 2, at a variable unit cost of $C_2(N)$ that depends on the number of jobs to be contracted out.
    *   Let the pdf of the weekly service demand be $f(N)$.
    *   An in-house crew member can process an average of m jobs per week.
        (a) Given the above information, construct an appropriate decision model.
        (b) Use the model to determine the optimal crew size when

$$C_w = 800, \ C_r = 50, \ C_1 = 75, \ m = 200$$

$$C_2(N) = \begin{cases} 90 - \frac{3}{20} \times N, & N < 200 \\ 60 & , \ N \geq 200 \end{cases}$$

$$f(N) = U[0, 2000]$$

Appendix 1: Statistics Primer

Once the equipment enters service a whole new set of information will come to light, and from this point on the maintenance program will evolve on the basis of data from actual operating experience. This process will continue throughout the service life of the equipment, so that at every stage maintenance decisions are based, not as an estimate of what reliability is likely to be, but on the specific reliability characteristics that can be determined at the time

F.S. Nowlan and H. Heap

A1.1 INTRODUCTION

Decisions relating to probabilistic maintenance problems, such as deciding when to perform preventive maintenance on equipment that is subject to breakdown, require information about when the equipment will reach a failed state. The engineer never knows exactly when the transition of the equipment from a good to a failed state will occur, but it is usually possible to obtain information about the probability of this transition occurring at any particular time. When optimal maintenance decisions are being determined, knowledge of statistics is required to deal with such probabilistic problems.

A1.2 RELATIVE FREQUENCY HISTOGRAM

If we think of a number of similar pieces of equipment that are subject to breakdown, we would not expect each of them to fail after the same number of operating hours. By noting the running time to failure of each item of equipment, it is possible to draw a histogram in which the area associated with any time interval shows the relative frequency of breakdown occurring in these intervals. This is illustrated in Figure A1.1. (In Appendix 2 we will deal with situations where there are not sufficient observations to construct a histogram, since we are dealing with a small size.)

If we now wish to determine the probability of a failure occurring between times t_{i-1} and t_i, we simply multiply the ordinate y by the interval (t_{i-1}, t_i). Further examination of Figure A1.1 will reveal that the probability of a failure occurring between t_0 and t_n, where t_0 and t_n are the earliest and latest times, respectively, at which the equipment has failed, is unity. That is, we are certain of the failure occurring in the interval (t_0, t_n), and the area of the histogram equals 1.

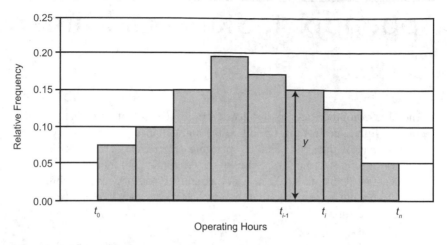

FIGURE A1.1 A histogram of time to failure.

A1.3 PROBABILITY DENSITY FUNCTION

In maintenance studies we tend to use probability density functions (pdf) rather than relative frequency histograms. This is because (1) the variable to be modeled, such as time to failure, is a continuous variable, (2) these functions are easier to manipulate, and (3) it should give a clearer understanding of the true failure distribution. Pdfs are similar to relative frequency histograms except that a continuous curve is used instead of bars, as shown in Figure A1.2. The equation of the curve of the probability density function is denoted by $f(t)$.

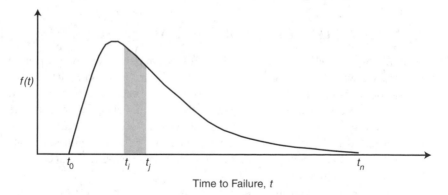

FIGURE A1.2 Probability density function.

For example, if we have $f(t) = 0.5 \exp(-0.5t)$, we get a curve of the shape of Figure A1.3. This is a probability density function of an exponential distribution.

Similar to the area under a relative frequency histogram, the area under the probability density curve also equals 1.

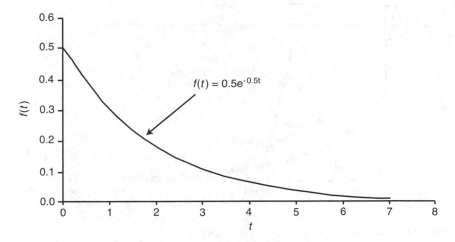

FIGURE A1.3 A probability density function of exponential distribution.

Referring back to Figure A1.2, the probability (risk) of a failure occurring between times t_i and t_j is the area of the shaded portion of the curve. Resorting to our knowledge of calculus, this area is the integral between t_i and t_j of $f(t)$, namely,

$$\int_{t_i}^{t_j} f(t)dt$$

The probability of a failure occurring between times t_0 and ∞ is then

$$\int_{t_0}^{\infty} f(t)dt = 1$$

Needless to say, the failure characteristics of different items of equipment are likely to be different from each other. Even the failure characteristics of identical equipment may not be the same if they are operating in different environments. There are a number of well-known pdfs that have been found in practice to describe the failure characteristics of equipment, and some of them are illustrated in Figure A1.4.

A1.3.1 HYPEREXPONENTIAL

When equipment has a failure time that can be very short or very long, its failure distribution can often be represented by the hyperexponential distribution. Some

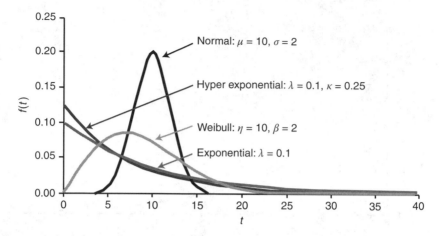

FIGURE A1.4 Common probability density functions.

computers have been found to fail according to this distribution. In the hyperexponential distribution, the short times to failure occur more often than in the negative exponential distribution, and similarly, the long times to failure occur more frequently than in the exponential case.

The density function of the hyperexponential distribution is

$$f(t) = 2k^2 \lambda \exp[-2k\lambda t] + 2(1-k)^2 \lambda \exp[-2(1-k)\lambda t]$$

for $t \geq 0$ with $0 < k \leq 0.5$, and where λ is the arrival rate of breakdowns and k is a parameter of the distribution.

A1.3.2 EXPONENTIAL

The exponential distribution is one that arises in practice where failure of equipment can be caused by failure of any one of a number of components of which the equipment is comprised. Also, it is characteristic of equipment that is subject to failure due to random causes, such as sudden excessive loading. The distribution is found to be typical for many electronic components and complex industrial plant.

The density function of the exponential distribution is

$$f(t) = \lambda \exp[-\lambda t] \qquad \text{for } t \geq 0$$

where λ is the arrival rate of breakdowns, and $1/\lambda$ is the mean of the distribution.

A probability function closely related to the exponential distribution is the Poisson distribution. If the time between failures of an item follows an exponential distribution, the arrival of failures is described as a Poisson process. The probability of observing n failures during the time interval $[0, t]$, $P_n(t)$, can be determined by the Poisson distribution, which has the following form:

$$P_n(t) = \frac{(\lambda t)^n \exp(-\lambda t)}{n!} \qquad \text{for } t \geq 0 \text{ and } n \text{ is a non-negative integer}$$

where λ is the mean arrival rate of failures.

A1.3.3 NORMAL

The normal (or Gaussian) distribution applies, for instance, when a random outcome (such as time to failure) is the additive effect of a large number of small and independent random variations. When this is true for the time to failure, the failure distribution is a bell-shaped normal function.

In practice, the lifetime of light bulbs and the time until the first failure of bus engines have been found to follow a normal distribution.

The density function of the normal distribution is

$$f(t) = \frac{1}{\sigma\sqrt{2\pi}} \exp\left[-\frac{1}{2}\left(\frac{t-\mu}{\sigma}\right)^2\right] \qquad \text{for } -\infty < t < \infty$$

where μ is the mean and σ the standard deviation of the distribution.

Note that for the normal distribution,

$$\int_0^\infty f(t)dt < 1 \qquad \text{but} \qquad \int_{-\infty}^\infty f(t)dt = 1$$

In practice, however, if the mean of the normal distribution, μ, is considerably removed from the origin $t = 0$ and the variance, σ^2, is not too large, then it is acceptable to use the normal distribution as an approximation to the real situation. A rough and ready rule would be that the mean μ should be greater than 3.5σ, since for this case there would be less than 1 chance in 4000 of the distribution giving a negative failure time.

A1.3.4 WEIBULL

The Weibull distribution fits a large number of failure characteristics of equipment. One of the original papers on the application of the Weibull distribution to equipment failure times was related to electron tubes.

The density function of the two-parameter Weibull distribution is

$$f(t) = \frac{\beta}{\eta}\left(\frac{t}{\eta}\right)^{\beta-1} \exp\left[-\left(\frac{t}{\eta}\right)^{\beta}\right] \qquad \text{for } t \geq 0$$

where η is the scale parameter (also known as the characteristic life), β is the shape parameter, and η and β are positive. When $\beta = 1$, the two-parameter Weibull is equivalent to the exponential distribution; when $\beta = 2$, it becomes the Rayleigh distribution. The Weibull approximates a normal distribution when, for example, $\beta = 3.44$.

A detailed discussion of the Weibull distribution is given in Appendix 2.

Before leaving pdfs, it should be noted that there are other distributions relevant to maintenance studies, including, for example, the gamma, Erlang, and lognormal. For the density functions of these distributions, and many others, the reader may refer to Evans et al. (2000).

A1.4   CUMULATIVE DISTRIBUTION FUNCTION

In maintenance studies we are often interested in the probability of a failure occurring before some specified time, say t. This probability can be obtained from the relevant probability density function as follows:

$$\text{Probability of failure before time } t = \int_{-\infty}^{t} f(t)dt$$

The integral $\int_{-\infty}^{t} f(t)dt$ is denoted by $F(t)$ and is termed the cumulative distribution function. As t approaches infinity, $F(t)$ tends to 1.

The form of $F(t)$ for the four density functions described in Section A1.3 is illustrated in Figure A1.5. $F(t)$ of a normal function can be obtained from the Standard Normal Distribution Table. This is explained in Table A1.1.

A1.5   RELIABILITY FUNCTION

A function complementary to the cumulative distribution function is the reliability function, also known as the survival function. It is determined from the probability that the equipment will survive at least to some specified time, say t. The reliability function is denoted by $R(t)$ and is defined as

$$R(t) = \int_{t}^{\infty} f(t)dt$$

and, of course, $R(t)$ also equals $1 - F(t)$. As t tends to infinity, $R(t)$ tends to zero.

The form of the reliability function for the four density functions described in A1.3 is illustrated in Figure A1.6.

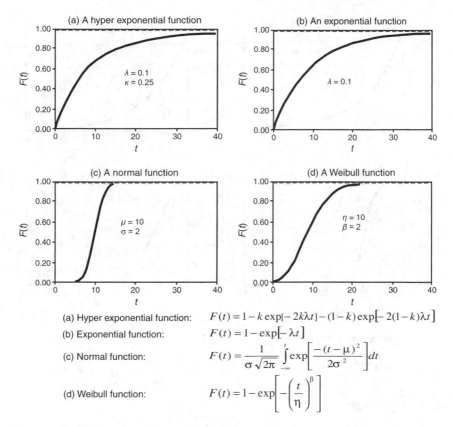

(a) Hyper exponential function: $\quad F(t) = 1 - k\exp[-2k\lambda t] - (1-k)\exp[-2(1-k)\lambda t]$

(b) Exponential function: $\quad F(t) = 1 - \exp[-\lambda t]$

(c) Normal function: $\quad F(t) = \dfrac{1}{\sigma\sqrt{2\pi}} \int_{-\infty}^{t} \exp\left[\dfrac{-(t-\mu)^2}{2\sigma^2}\right] dt$

(d) Weibull function: $\quad F(t) = 1 - \exp\left[-\left(\dfrac{t}{\eta}\right)^{\beta}\right]$

FIGURE A1.5 Cumulative distribution function.

TABLE A1.1
Standard Normal Distribution Table

The cumulative distribution function, $F(t)$, of a normal function with mean $= \mu$ and standard deviation $= \sigma$ can be determined from the standard normal distribution table given in Appendix 6, which tabulates the value of $1 - \Phi(z)$, where z ($= (t - \mu)/\sigma$) is a standardized normal variable and $\Phi(z)$ is the cumulative distribution function of the standard normal distribution. Thus, the table provides the probability that the standardized normal variable chosen at random is greater than a specified value of z.

The normal distribution being symmetrical about its mean, $\Phi(-z) = 1 - \Phi(z)$. Thus, only the probability for $z \geq 0$ is tabulated.

Consider an item that is operational at time t_1 when a mission starts. We may wish to determine the probability of the item surviving the mission of duration t. The required measure can be expressed in the usual notation of conditional probability as

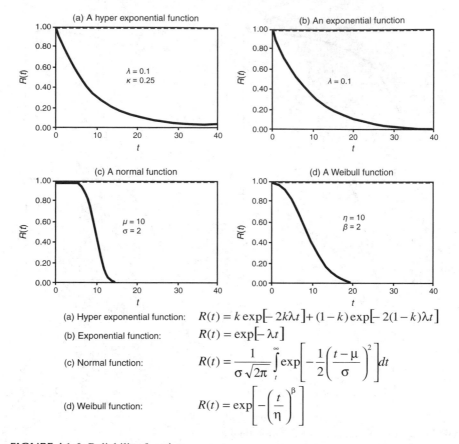

FIGURE A1.6 Reliability function.

$$R(t_1 + t \mid t_1) = P(T \ge t_1 + t \mid T \ge t_1) = \frac{P(T \ge t_1 + t)}{P(T \ge t_1)} = \frac{R(t_1 + t)}{R(t_1)} = \frac{\int_{t_1+t}^{\infty} f(t)dt}{\int_{t_1}^{\infty} f(t)dt} \quad \text{(A1.1)}$$

where T is time to failure.

If the failure time follows an exponential distribution, as shown in Figure A1.7, Equation A1.1 will become

$$R(t_1 + t \mid t_1) = \frac{\int_{t_1+t}^{\infty} \lambda \exp(-\lambda t)dt}{\int_{t_1}^{\infty} \lambda \exp(-\lambda t)dt} = \frac{\exp\left[-\lambda(t_1 + t)\right]}{\exp\left[-\lambda t_1\right]} = \exp(-\lambda t) = R(t)$$

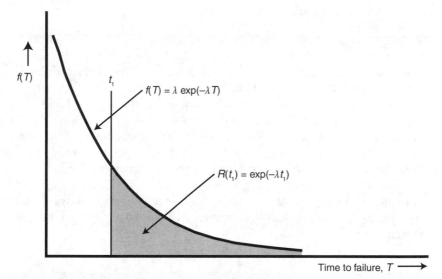

FIGURE A1.7a Exponential distribution: reliability at t_1.

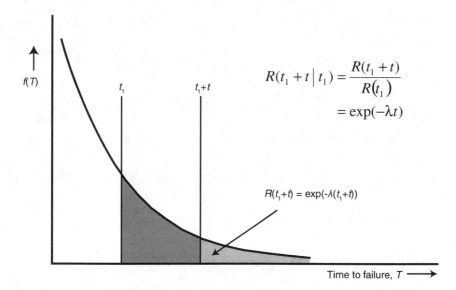

FIGURE A1.7b Exponential distribution: reliability at $t_1 + t$, given the item is operational at t_1.

Thus, for operational items with failure times that are exponentially distributed, $R(t_1 + t \mid t_1) = R(t)$. In other words, their chance of survival (or conversely, their risk of failure) in the next instance is independent of their current age. This memoryless property is unique to the exponential distribution, the only continuous distribution with this feature.

A1.6   HAZARD RATE

A statistical characteristic of equipment frequently used in replacement studies is the hazard rate.

   To introduce the hazard rate, consider a test where a large number of identical components are put into operation and the time to failure of each component is noted. An estimate of the hazard rate of a component at any point in time may be thought of as the ratio of a number of items that failed in an interval of time (say, 1 week) to the number of items in the original population that were operational at the start of the interval. Thus, the hazard rate of an item at time t is the probability that the item will fail in the next interval of time given that it is good at the start of the interval; that is, it is a conditional probability.

   Specifically, letting $h(t)\,\delta t$ be the probability that an item fails during a short interval δt, given that it has survived to time t, the usual notation for conditional probability may be written as

$P(A|B)$ = probability of event A occurring once it is known that B has occurred

$\quad = h(t)\,\delta t$

where A is the event "failure occurs in interval δt" and B is the event "no failure has occurred up to time t."

$$P(A|B) \text{ is given by}$$

$$P(A|B) = \frac{P(A \text{ and } B)}{P(B)}$$

where $P(A \text{ and } B)$ is the probability of both events A and B occurring and

$$P(A \text{ and } B) = \int_{t}^{t+\delta t} f(t)dt$$

$P(B)$ is the probability of event B occurring and

$$P(B) = \int_{t}^{\infty} f(t)dt$$

Therefore, the hazard rate in interval δt is

$$h(t)\delta t = \frac{\displaystyle\int_{t}^{t+\delta t} f(t)dt}{\displaystyle\int_{t}^{\infty} f(t)dt} = \frac{F(t+\delta t)-F(t)}{1-F(t)} \qquad \text{(A1.2)}$$

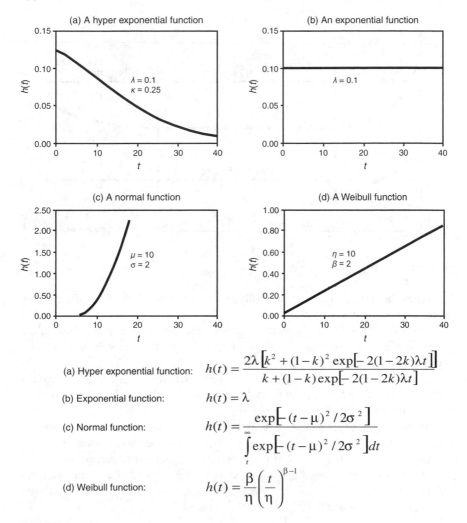

(a) A hyper exponential function

$\lambda = 0.1$
$\kappa = 0.25$

(b) An exponential function

$\lambda = 0.1$

(c) A normal function

$\mu = 10$
$\sigma = 2$

(d) A Weibull function

$\eta = 10$
$\beta = 2$

(a) Hyper exponential function: $h(t) = \dfrac{2\lambda\left[k^2 + (1-k)^2\,\exp\left[-2(1-2k)\lambda t\right]\right]}{k + (1-k)\exp\left[-2(1-2k)\lambda t\right]}$

(b) Exponential function: $h(t) = \lambda$

(c) Normal function: $h(t) = \dfrac{\exp\left[-(t-\mu)^2/2\sigma^2\right]}{\displaystyle\int_{t}^{\infty}\exp\left[-(t-\mu)^2/2\sigma^2\right]dt}$

(d) Weibull function: $h(t) = \dfrac{\beta}{\eta}\left(\dfrac{t}{\eta}\right)^{\beta-1}$

FIGURE A1.8 Hazard rate.

If Equation A1.2 is divided through by δt, and then $\delta t \to 0$, this gives

$h(t) = \dfrac{f(t)}{1 - F(t)}$, where $h(t)$ is termed hazard rate, also known as instantaneous

failure rate.

The form of the hazard rate for the distributions discussed in Section A1.3 is illustrated in Figure A1.8.

An interesting point to note about the hyperexponential distribution is that as time increases, the hazard rate decreases. This may be interpreted as an improvement in the equipment over time and may be the case with equipment that requires small adjustments after an overhaul or replacement to get it completely operational. Equipment that work hardens can also be modeled by this distribution.

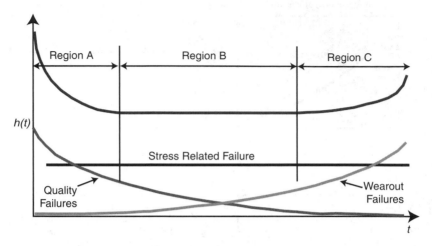

FIGURE A1.9 Bathtub curve.

When the hazard rate increases with time, such as for the normal distribution, it indicates an aging or wear-out effect.

With the exponential distribution, the hazard rate is constant. This failure pattern can be the result of completely random events such as sudden stresses and extreme conditions.

It also applies to the steady-state condition of complex equipment that fails when any one of a number of independent constituent components breaks, or any one of a number of failure modes occurs.

Before leaving this aspect, it is interesting to note the form, illustrated in Figure A1.9, that the hazard rate sometimes takes with complex equipment. For obvious reasons, such a pattern is often referred to as the bathtub curve.

The bathtub curve may be interpreted as the aggregated effect of three categories of failures: quality failures, stress-related failures, and wear-out failures.

Regions A, B, and C of Figure A1.9 are labeled as:

A = A running-in period
B = Normal operation where failures that occur are predominantly due to chance
C = Deterioration, i.e., due to wear-out

A common problem in maintenance is to determine the most appropriate policy to adopt when equipment is in one of regions A, B, and C. If the only form of maintenance possible is replacement, either on a preventive basis or because of failure, then in regions A and B no preventive replacements should occur since such replacements will not reduce the risk of equipment failure. If preventive replacements are made in regions A and B, then maintenance effort is being wasted. Unfortunately, this is often the case in practice, since it is often mistakenly assumed that as equipment ages, the risk of failure will increase. In region C, preventive replacement will reduce the risk of equipment failure in the future, and just when these preventive replacements should occur will be influenced by the relative costs or other relevant impact factors, such as downtime of preventive and failure replacements. Such replacement problems are covered in Chapter 2.

When maintenance policies are more general than only replacement, such as including an overhaul that may not return the equipment to a statistically as-good-as-new condition, then preventive maintenance may be worthwhile in all three regions. Such policies are discussed in Chapter 3.

REFERENCE

Evans, M., Hastings, N., and Peacock, B. (2000). *Statistical Distributions*, 3rd ed. New York: Wiley.

FURTHER READING

Hines, W.W., Montgomery, D.C., Goldsman, D.M., and Borror, C.H. (2003). *Probability and Statistics in Engineering*, 4th ed. New York: Wiley.

Murdoch, J. and Barnes, J.A. (1970). *Statistical Tables for Science, Engineering, Management and Business Studies*, 2nd ed. New York: Macmillan.

Walpole, R.E., Myers, R.H., Myers, S.L., and Ye, K. (2002). *Probability and Statistics for Engineers and Scientists*, 7th ed. Englewood Cliffs, NJ: Prentice Hall.

PROBLEMS

1. The failure times for a model 555 rifle has demonstrated a normal probability density function with $\mu = 100$ hours and $\sigma = 10$ hours. Find the reliability of such a rifle for a mission time of 104 hours and the hazard rate of one of these rifles at age 105 hours.

2. A computer has a constant error rate of one error every 17 days of continuous operation. What is the reliability associated with the computer to correctly solve a problem that requires 5 hours time? 25 hours time?

3. The failure times of JP29M transmitting tubes have a Weibull distribution with $\beta = 2$ and $\eta = 1000$ hours. Find the reliability of one of these tubes for a mission time of 100 hours and the hazard rate associated with one that has operated successfully for 100 hours.

4. Jackleg drill failures have been analyzed, and the failure distribution is found to be uniform within the interval 0 to 2000 hours of operation. That is, the Pdf of the failure distribution has a constant value within the specified interval and 0 elsewhere. What is the probability of a drill continuing to operate satisfactorily for a project period of 20 hours given that the drill had already been used for 1200 hours?

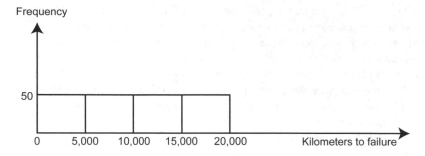

FIGURE A1.10 Water pump failure pattern.

5. Failure times of a type GLN microwave tube have been observed to follow a normal distribution with $\mu = 5000$ hours and $\sigma = 1000$ hours. Find the reliability of such a tube for a mission time of 4100 hours and the hazard rate of one of these tubes at age 4400 hours.

6. A component's constant hazard rate is 20 failures/10^6 hours.
   Write down its failure density function and sketch its form.
   Write down its hazard rate function and sketch its form.
   Write down its reliability function for mission t and sketch its form.
   Write down its reliability function for a mission t, starting the mission at age T.

7. Truck water pump failures have been analyzed, and it is found through using a statistical goodness-of-fit test that the pump failure times can be described adequately by the uniform distribution shown in Figure A1.10. Given this failure pattern, plot to scale:

   $f(t)$, the probability density function
   $R(t)$, the reliability function
   $h(t)$, the hazard rate function

Appendix 2: Weibull Analysis

Weibull analysis is the world's most popular method of analyzing and predicting failures and malfunctions of all types. The method identifies the category of failure: infant mortality, random or wear out. Weibull analysis provides the quantitative information needed for making RCM decisions which are often made from a qualitative approach

Paul Barringer

A2.1 WEIBULL DISTRIBUTION

The Weibull distribution is named after Waloddi Weibull (1887–1979), who found that, in general, distributions of data on product life can be modeled by a function of the following form:

$$f(t) = \frac{\beta}{\eta}\left(\frac{t-\gamma}{\eta}\right)^{\beta-1} \exp\left(-\left(\frac{t-\gamma}{\eta}\right)^{\beta}\right) \qquad \text{for } t > \gamma$$

$$f(t) = 0 \qquad \text{for } t \leq \gamma$$

The three parameters of a Weibull distribution are β (the shape parameter), γ (the location parameter), and η (the scale parameter). β and η are greater than 0.

Consider the case when $\gamma = 0$ (which is usually the case when dealing with component preventive replacement — see Chapter 2) and η is kept constant; Weibull distributions for $\beta = 0.5, 1, 2.5, 3.44,$ and 5 are shown in Figure A2.1.

A2.1.1 Shape Parameter

The β value determines the shape of the distribution. When $\beta < 1$, the Weibull distribution has a hyperbolic shape with $f(0) = \infty$. When $\beta = 1$, it becomes an exponential function. When β exceeds 1, it is a unimodal function where skewness changes from left to right as the value of β increases. When $\beta \approx 3.44$, the Weibull distribution approximates the symmetrical normal function. Hence, β is termed the shape parameter.

The hazard rate, $h(t)$, of the Weibull distribution is of the following form:

$$h(t) = \frac{\beta}{\eta}\left(\frac{t-\gamma}{\eta}\right)^{\beta-1} \qquad \text{when } t > \gamma$$

$$= 0 \qquad \text{otherwise}$$

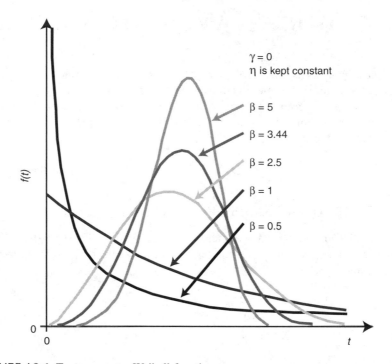

FIGURE A2.1 Two-parameter Weibull functions.

Clearly, $h(t)$ varies with the value of the independent variable t, as shown in Figure A2.2. In particular, when $\beta < 1$, $h(t)$ is a decreasing function of t. When $\beta = 1$, $h(t)$ does not vary with t; $h(t)$ becomes an increasing function of t when $\beta > 1$.

A2.1.2 SCALE PARAMETER

Figure A2.3 shows two Weibull distributions, both with identical γ and β values, but different in their η values. While both share the same shape, the spread of these distributions is proportional to the η value. Hence, η is termed the scale parameter.

The cumulative distribution function, $F(t)$, of the Weibull distribution is

$$F(t) = 1 - \exp\left(-\left(\frac{t-\gamma}{\eta}\right)^{\beta}\right)$$

When $t - \gamma = \eta$, $F(t) = 1 - \exp(-1)$, or approximately 63.2%, for all values of β. Thus, η is also known as the characteristic life of the Weibull distribution.

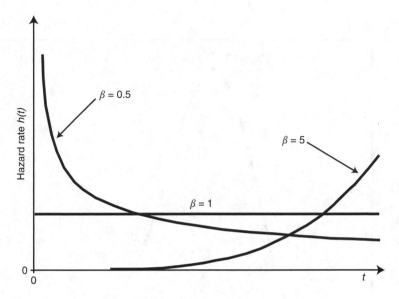

FIGURE A2.2 Hazard rate of Weibull distribution.

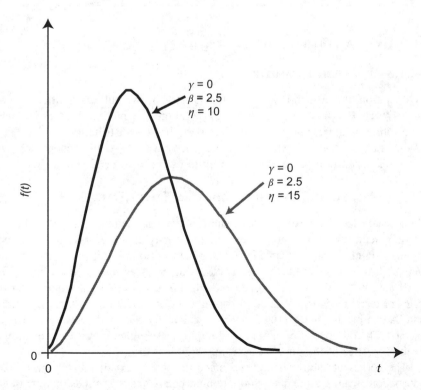

FIGURE A2.3 Two Weibull distributions with identical location and shape parameters but different scale parameters.

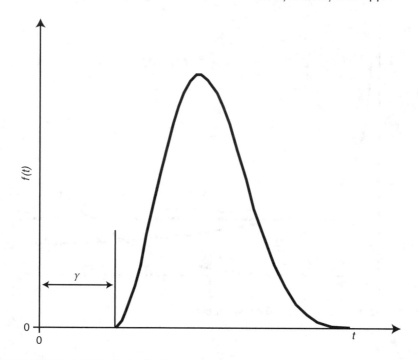

FIGURE A2.4 A Weibull distribution with $\gamma > 0$.

A2.1.3 LOCATION PARAMETER

By definition, the probability density function of the Weibull distribution is zero for $t \leq \gamma$. That is, there is no risk of failure before γ, which is therefore termed the location parameter or the failure-free period of the distribution.

In practice, γ may be negative, in which case the equipment may have undergone a run-in process or it had been in use prior to $t = 0$ (Figure A2.4).

A2.1.4 FITTING A DISTRIBUTION MODEL TO SAMPLE DATA

Maintenance decision analysis often requires the use of the failure time distribution of equipment, which may not be known. There may, however, be a set of observations of failure times available from historical records. We might wish to find the Weibull distribution that fits the observations, and to assess the goodness of the fit.

If data in the form of historical records are not available, a specific test or series of tests needs to be made to obtain a set of observations, that is, sample data. A sample is characterized by its size and by the method by which it is selected. The purpose of obtaining the sample is to enable inferences to be drawn about properties of the population from which it is drawn.

The comments above have been made in the context of identifying failure distributions from sample data. Similar points can be made about estimating trend lines from sample data, such as the trend in equipment operating costs.

Techniques available for identifying probability distributions from a sample are discussed in the subsequent sections of this appendix.

A2.2   WEIBULL PAPER

The form of the cumulative distribution function, $F(t)$, of a Weibull distribution for $\gamma = 0$ is

$$F(t) = 1 - \exp\left(-\left(\frac{t}{\eta}\right)^{\beta}\right)$$

With simple manipulation, the above equation can be transformed into the linear expression

$$\ln \ln\left(\frac{1}{1-F(t)}\right) = \beta \ln t - \beta \ln \eta$$

Thus, a plot of $\ln \ln\left(\dfrac{1}{1-F(t)}\right)$ vs. $\ln t$ will give a straight line when t is generated from a Weibull distribution with $\gamma = 0$. Special graph paper, known as Weibull paper, with the vertical axis in ln ln scale and the horizontal axis in ln scale, makes it possible to plot $F(t)$ and t directly. Figure A2.5 shows two-cycle Weibull paper — paper with an abscissa scale that spreads over a range of 10^2 units of the life value (Nelson, 1967).

A2.3   WEIBULL PLOT

A2.3.1   Estimating Cumulative Percent Failure, $F(t)$

Plotting failure data on Weibull paper involves estimation of $F(t_i)$ for every observed failure time t_i. Consider five failures at 2, 7, 13, 19, and 27 cycles. Let i denote the rank of an observation when data are sorted in ascending order. In this example, $i = 1$ for 2 cycles, and $i = 5$ for 27 cycles. Using i/n as an estimate, the values of $F(t_i)$ for the sampled data are 20, 40, 60, 80, and 100%. That is, 100% of the items are expected to fail at 27 cycles. Obviously, the $F(t_i)$ values thus determined are pessimistic estimates.

A better estimate of $F(t_i)$ is to use the median rank table given in Appendix 8. The determination of median ranks is explained in Table A2.1. The first row of a median rank table shows the sample size n, and the first column indicates the rank number i. For a sample of five observations, the values of $F(t_i)$ are 12.9, 31.4, 50.0, 68.6, and 87.1%. Using this method, the chance of these estimates

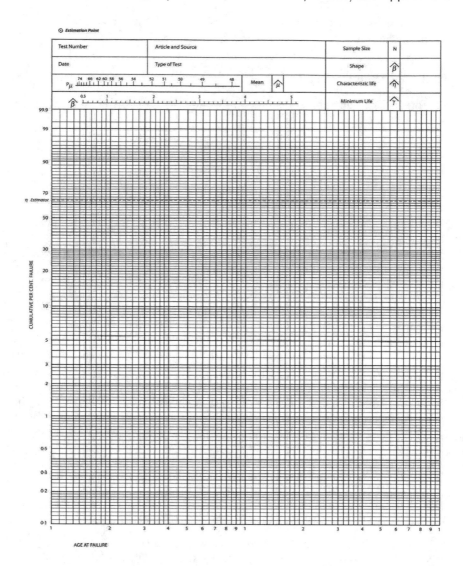

FIGURE A2.5 Two-cycle Weibull paper.

being optimistic is equal to that of them being pessimistic. The concept of median rank is also explained in the WeibullSoft package that can be downloaded from http://www.crcpress.com/e_products/downloads/download.asp?cat_no=DK9669.

For sample sizes greater than 12 but less than 100, Benard's approximation for the median rank is adequate as an estimate of $F(t_i)$:

$$F(t_i) \approx \frac{i - 0.3}{n + 0.4} \ .$$

TABLE A2.1
Median Ranks

The median rank is the solution for $F(t)$ in the following equation:

$$\sum_{r=i}^{n} \frac{n!}{r!(n-r)!} \left[F(t) \right]^r \left[1-F(t) \right]^{n-r} = 0.5 \qquad \text{(A2.1)}$$

where i is the ranked-order number of our observation and n is the sample size. The left-hand side of Equation A2.1 evaluates the probability of observing i or more failures at time t in n observations.

Example:
Suppose we have observed 10 items and the median rank of the third-ranked failure at time t is to be determined. In this case, $i = 3$, $n = 10$, we solve for $F(t)$ in Equation A2.1.

$$\sum_{r=3}^{10} \frac{10!}{3!7!} \left[F(t) \right]^r \left[1-F(t) \right]^{10-r} = 0.5$$

Thus, the median rank of the third-ranked failure is 0.25857.

Solving for $F(t)$ in Equation A2.1 involves the use of cumulative binomial probabilities tables (Murdoch and Barnes, 1970). This can be a tedious process. However, it can be simplified by using the median rank table in Appendix 8.

For sample sizes greater than 100, the effects of small sample bias are insignificant and $F(t_i)$ may be estimated from the expression for mean ranks:

$$F(t_i) \approx \frac{i}{n+1}$$

A2.3.2 ESTIMATING THE PARAMETERS

The procedure for fitting a Weibull distribution to a data set of failure times is explained with the use of a worked example relating to lamp failures. Since there are numerous failure observations in the example, data are grouped into a number of nonoverlapping intervals of failure time, as shown in Table A2.2. The cumulative probability of failure $F(t_i)$ for the end of each time interval is equal to the cumulative number of failures observed up to the end of the interval divided by the original number of lamps in the sample. A Weibull plot of the data set is given in Figure A2.6, in which the value of $F(t_i)$ is plotted at the time corresponding to the end of the interval (t_i) because it is the cumulative value for the interval $[0, t_i]$.

If we can fit a straight line through the Weibull plot, such as the case shown in Figure A2.6, the Weibull distribution with $\gamma = 0$ can be used as the model of

TABLE A2.2
Lamp Failure Data

End of Time Interval, $< t_i$	Cumulative Probability, $F(t_i)$ (%)
< 04	5
< 08	14
< 12	20
< 16	25
< 20	32
< 24	38
< 28	46
< 32	48
< 36	54
< 40	60
< 44	64
< 48	66
< 52	
< 56	
< 60	78
< 64	
< 68	
< 72	
< 76	
< 80	86

the data set. We can then proceed to estimate the other parameters of the distribution from the plot.

From the estimation point on the top left-hand corner of the Weibull paper, we draw a line perpendicular to the fitted line. The intersection between the perpendicular line and the $\hat{\beta}$ scale beneath the estimation point gives the estimated value of β. The value of t at which the fitted line cuts $F(t) = 63.2\%$ (the η estimation line on Weibull paper) is an estimate of η.

While a Weibull distribution is completely defined by the values of its γ, β, and η parameters, we may also wish to determine its mean value μ. It can also be determined from the Weibull plot, from the intersection of the perpendicular line and the P_μ scale beneath the estimation point of the Weibull paper. In the example given in Figure A2.6, $P_\mu = 60\%$. Thus, the distribution mean is 40 hours, the time at which the cumulative probability of failure is 60%.

Both the mean μ and standard deviation σ of the Weibull distribution can also be determined analytically using the following expressions:

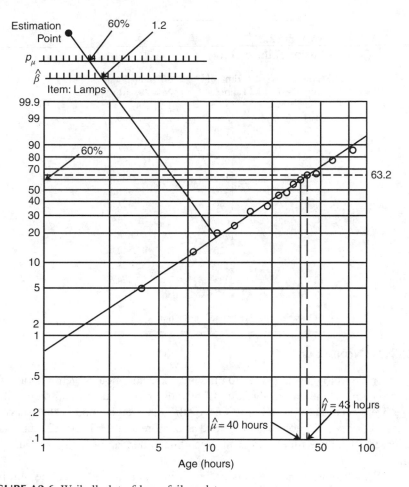

FIGURE A2.6 Weibull plot of lamp failure data.

$$\mu = \eta \Gamma \left(1 + \frac{1}{\beta} \right) + \gamma$$

$$\sigma^2 = \eta^2 \left[\Gamma \left(1 + \frac{2}{\beta} \right) - \Gamma^2 \left(1 + \frac{1}{\beta} \right) \right]$$

where $\quad \Gamma(z) = \displaystyle\int_{0}^{\infty} t^{z-1} e^{-t} dt \quad$ and $\quad \Gamma(z) = (z-1)\Gamma(z-1)$

Values of the gamma function, $\Gamma(z)$, for $z = 1$ to 2 are given in Appendix 7.

TABLE A2.3
Motor Failure Data

Failure Number, i	Time of Failure, t_i	Median Ranks, $N = 20$, $F(t_i)$ (%)
1	550	3.406
2	720	8.251
3	880	13.147
4	1020	18.055
5	1180	22.967
6	1330	27.880
7	1490	32.975
8	1610	37.710
9	1750	42.626
10	1920	47.542
11	2150	52.458
12	2325	57.374
13–20	Censored data[a]	

[a] Censored data are defined in Section A2.7.

A2.3.3 Nonlinear Plot

Consider the following example: 20 randomly selected motors were put on a life test program. At the end of the test, 12 of these motors failed and their failure times were recorded. The failure times of the other eight motors that survived the test (i.e., $t_i > 2325$ hours) are regarded as censored data.* Table A2.3 tabulates the observed failure times sorted in ascending order, along with their median ranks.

A Weibull plot of these results is shown in Figure A2.7. Since the plot is a curve, we have to use a three-parameter Weibull distribution to model the data set. The curvature of the plot suggests that the location parameter γ is >0. Obviously, γ must be less than or equal to the shortest failure time, t_1. Finding the correct value of γ will produce a linear plot.

We can find γ by trial and error, subtracting different values of γ ($0 \le \gamma \le t_1$) from every t_i until we obtain a straight line for the plot of ($t_i - \gamma$) vs. $F(t_i)$ on Weibull paper. In this example, a good straight line can be obtained for $\gamma = 375$ hours. The adjusted failure times, ($t_i - \gamma$), are tabulated in Table A2.4. The Weibull plot of the adjusted data is a straight line, as shown in Figure A2.8.

Summarizing the results, the parameters of the fitted distribution are $\gamma = 375$ hours, $\beta = 1.32$, and $\eta = 2120$ hours.

The probability density function of the fitted distribution is shown in Figure A2.9.

---

* Censored data are defined in Section A2.7.

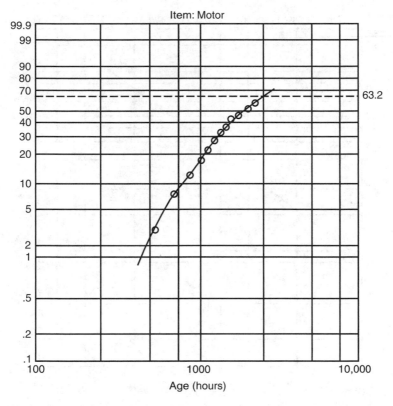

FIGURE A2.7 Weibull plot of motor failure data.

TABLE A2.4
Adjusted Motor Failure Data, $t_i - \hat{\gamma}$

Failure Number, i	Adjusted Failure Data, $t_i - \hat{\gamma}$	Median Ranks, $N = 20$, $F(t_i)$ (%)
1	175	3.406
2	345	8.251
3	505	13.147
4	645	18.055
5	805	22.967
6	955	27.880
7	1115	32.975
8	1235	37.710
9	1375	42.626
10	1545	47.542
11	1775	52.458
12	1950	57.374
13–20	Censored data	

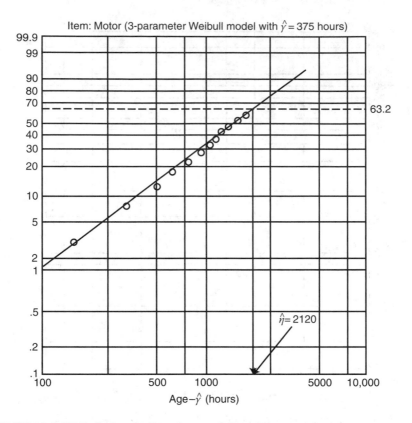

FIGURE A2.8 Weibull plot of adjusted motor failure data.

Apart from the trial-and-error approach, there is a more direct way to obtain the γ value. It is a graphical method involving the following steps:

1. Select 2 endpoints of the Weibull plot that cover the entire set of failure data. Let a and c be the projections of these endpoints on the $F(t)$-axis, as shown in Figure A2.10.
2. Bisect the distance between a and c. Let b be the mid-point.
3. Let the projections of a, b, and c on the t-axis be t_1, t_2, and t_3, respectively.
4. The location parameter γ can be estimated from the following equation:

$$\hat{\gamma} = t_2 - \frac{(t_3 - t_2)(t_2 - t_1)}{(t_3 - t_2) - (t_2 - t_1)}$$

We will now use the graphical method to find the location parameter. From the Weibull plot of Figure A2.11, we have $t_1 = 500$ hours, $t_2 = 933$ hours, and $t_3 = 2500$ hours.

$$f(t) = \frac{1.32}{2120}\left(\frac{t-375}{2120}\right)^{0.32} \exp\left(-\left(\frac{t-375}{2120}\right)^{1.32}\right) \qquad \text{for } t \geq 375$$

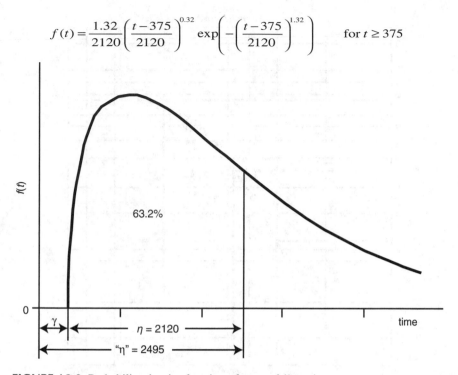

FIGURE A2.9 Probability density function of motor failure time.

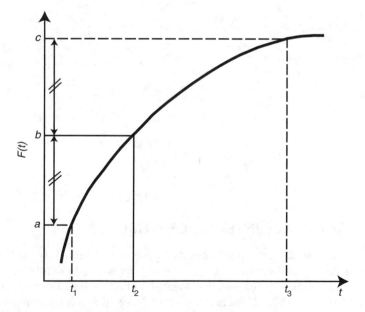

FIGURE A2.10 A nonlinear Weibull plot.

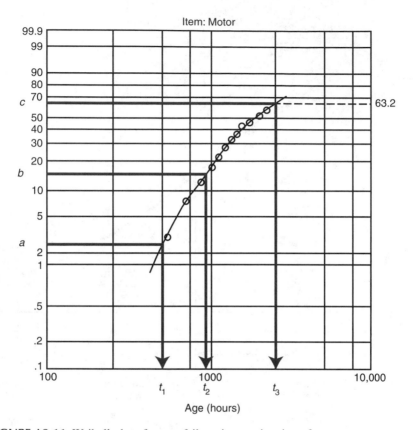

FIGURE A2.11 Weibull plot of motor failure time: estimation of γ.

Using the equation

$$\hat{\gamma} = t_2 - \frac{(t_3 - t_2)(t_2 - t_1)}{(t_3 - t_2) - (t_2 - t_1)}$$

$$\hat{\gamma} = 933 - \frac{(2500 - 933)(933 - 500)}{(2500 - 933) - (933 - 500)}$$

$$= 335 \text{ hours}$$

A2.4   CONFIDENCE INTERVAL OF WEIBULL PLOT

We analyze failure data in order to estimate measures of reliability such as $F(t)$ from the Weibull plot. For more security, we will determine a confidence interval on our estimation of $F(t)$. Suppose we want to find a $(1 - \alpha)$ confidence interval for $F(t)$, where α is the risk we are willing to accept that the interval we found does not contain the true $F(t)$. For example, when we establish a 90% confidence

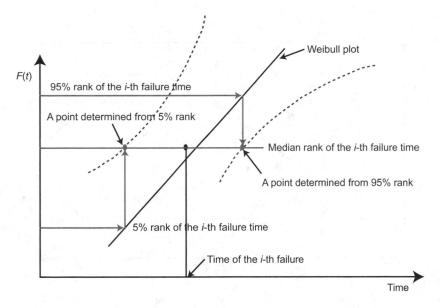

FIGURE A2.12 Determining confidence interval of a Weibull plot: method 1.

interval for $F(t)$, it means that we are 90% confident that the real $F(t)$ is contained in the confidence interval.

To build the confidence interval, we need to use 5% and 95% ranks. Tables of 5% and 95% ranks are given in Appendices 9 and 10.

The procedure for determining the 90% confidence limits on $F(t)$ of a failure time is illustrated in Figure A2.12.

There is an alternative procedure for constructing the confidence interval, as shown in Figure A2.13. In this alternative procedure, the three points corresponding to 5%, 50%, and 95% ranks of a given failure time t are plotted on a vertical line rather than a horizontal line (O'Connor, 2002). With this procedure, however, we often cannot use it to determine the lower bound for the confidence interval of the B_{10} life without extrapolation if the size of the data set used to create the Weibull plot is small (the B_{10} life will be introduced in Section A2.5). The procedure illustrated in Figure A2.12 does not have this problem.

EXAMPLE

We tested 10 electrical batteries for 9 hours. At the end of the test, two were still working. Here are the times when the other eight failed: 1.25, 2.40, 3.20, 4.50, 5.00, 6.50, 7.00, and 8.25 hours.

The data for establishing the 90% confidence interval on the estimate of $F(t)$ are given in Table A2.5.

Using the procedure explained in Figure A2.12, the Weibull plot with the 90% confidence interval is shown in Figure A2.14, from which we can say with 90% confidence that at $t = 4.5$ hours the cumulative distribution

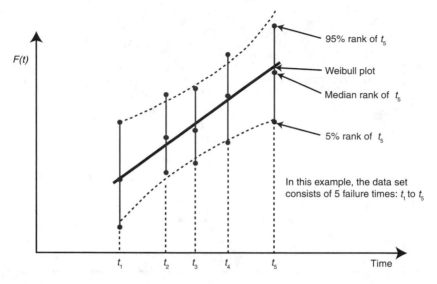

FIGURE A2.13 Determining confidence interval of a Weibull plot: method 2.

TABLE A2.5
Battery Failure Data: Median, 5%, and 95% Ranks

Failure Number	t_i	Median Ranks	5% Ranks	95% Ranks
1	1.25	6.697	0.512	25.887
2	2.40	16.226	3.677	39.416
3	3.20	25.857	8.726	50.690
4	4.50	35.510	15.003	60.662
5	5.00	45.169	22.244	69.646
6	6.50	54.831	30.354	77.756
7	7.00	64.490	39.338	84.997
8	8.25	74.142	49.310	91.274
Appendix used		8	9	10

function $F(t)$ will have a value between 16% and 64%. In other words, after
4.5 hours, in 90% of similar tests between 16% and 64% of the batteries
will have stopped working.

If we want a 90% confidence interval on the reliability $R(t)$ at time 4.5
hours, we take the complement of the limits on the confidence interval for
$F(t)$, that is, $(100 - 64, 100 - 16)$. In other words, we are 90% confident
that reliability at time $t = 4.5$ hours is between 36% and 84%. Or we can
say that we are 95% confident that the reliability after 4.5 hours of operation
will not be less than 36%.

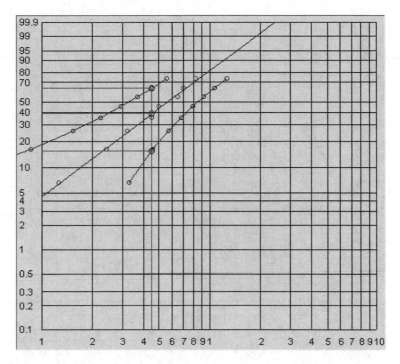

FIGURE A2.14 Weibull plot of battery failure time: 90% confidence interval of $F(t)$.

A2.5 B_q LIFE

The B_q life of an item is the time at which $q\%$ of the population will have failed. To illustrate, let us take the example in Section A2.4. From Figure A2.15, we obtain the following results:

The point estimate of B_{20} life is 2.72 hours. The 90% confidence interval on the B_{20} life is between 1.1 hours and 5.0 hours. B_{10} is often quoted in the specification of bearing life.

A2.6 KOLMOGOROV–SMIRNOV GOODNESS-OF-FIT TEST

The Kolmogorov–Smirnov (K-S) test is an appropriate tool to determine if a hypothesized distribution fits a data set. The test can be used for small as well as large sample sizes. It is limited, though, to the evaluation of hypothesized distributions that are continuous and completely specified; for example, t is exponentially distributed with $\lambda = 10$, or t is generated from a Weibull distribution with $\gamma = 0$, $\beta = 2$, and $\eta = 2150$.

The K-S procedure tests the hypothesis that the cumulative distribution function, $F_0(t)$, is $F(t)$. A random sample of size n is drawn from a continuous

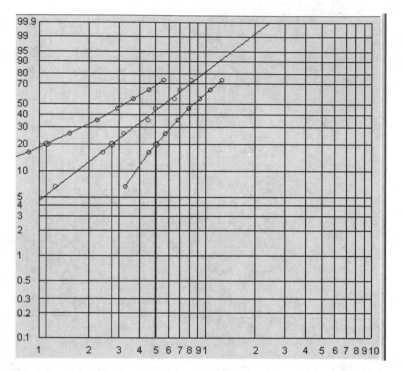

FIGURE A2.15 Weibull plot of battery failure time: 90% confidence interval of B_{20}.

distribution $F(t)$. Let the sample cumulative distribution function be $\hat{F}(t)$, and it is estimated by median rank (for small n) or mean rank (for large n). $\hat{F}(t)$ is then compared with the hypothesized $F(t)$. If $\hat{F}(t)$ deviates too much from $F(t)$, the hypothesis that $F_0(t) = F(t)$ is rejected.

Suppose we want to test the hypothesis at a significance level of α; this means that we are willing to accept α as the risk of wrongly rejecting the hypothesis, that is, $F_0(t) = F(t)$, when it is true. This is known as type I error in statistical tests. Reducing the significance level, α, without increasing the sample size at the same time will increase another type of risk — the risk of wrongly accepting the hypothesis when it is not true; this is known as a type II error. The significance level used represents a tradeoff between the cost of sampling and the risk of making wrong decisions from the test. Typical values of significance level are 1%, 2%, 5%, 10%, and 20%.

The K-S test statistic is $d = \max_i \left| F(t_i) - \hat{F}(t_i) \right|$.

When the hypothesis that $F_0(t) = F(t)$ is true, d has a distribution that is a function of n but which is independent of $F_0(t)$. The hypothesis that $F_0(t) = F(t)$ is rejected at the α level of significance whenever $d > d_\alpha$. The values of d_α are given in Appendix 11.

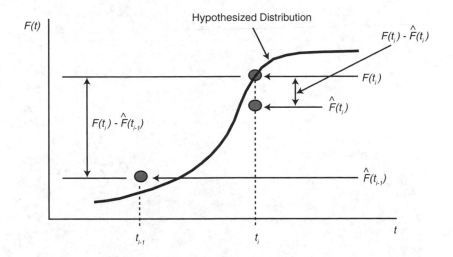

FIGURE A2.16 Kolmogorov–Smirnov goodness-of-fit test.

In principle, the differences between $\hat{F}(t_i)$ and $F(t_i)$ must be examined for all t_i. In reality, these differences need be examined only at the jump points of $F(t)$. The jump points occur at the observed values of the random variables. At each jump point t_i the two differences $\left|F(t_i) - \hat{F}(t_i)\right|$ and $\left|F(t_i) - \hat{F}(t_{i-1})\right|$ must be obtained (see Figure A2.16). For a random sample of size n, the maximum absolute deviation, d, must be among these $2n$ differences. Hence,

$$d = \max_{i} \left(\left|F(t_i) - \hat{F}(t_i)\right|, \left|F(t_i) - \hat{F}(t_{i-1})\right| \right)$$

EXAMPLE

We have tested five items to failure and the failure times are 2, 5, 6, 8, and 10 hours.

Using the WeibullSoft package, which can be downloaded from the Web site http://www.crcpress/mrr, the estimated parameters of the Weibull distribution that fits the data set are determined to be $\gamma = 0$, $\beta = 1.64$, and $\eta = 7.36$ hours.

Figure A2.17 is a screen dump of the output of the package when the data set is analyzed. Hitting the "goodness-of-fit test (K-S test)" button causes the K-S test summary window to pop up. The window indicates that the test statistic $d = 0.283$, and the critical value $d_\alpha = 0.51$ at 10% significance level. Since $d < d_\alpha$, the fitted Weibull distribution is not rejected as a model of the data set.

The calculations involved in the test are shown in Table A2.6.

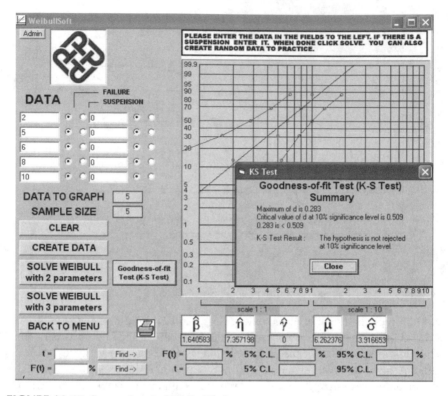

FIGURE A2.17 Screen dump of WeibullSoft.

TABLE A2.6
K-S Test: Determining the Test Statistic, *d*

t_i	$F(t_i)$[a]	$\hat{F}(t_i)$[b]	$\left|F(t_i) - \hat{F}(t_i)\right|$	$\left|F(t_i) - \hat{F}(t_{i-1})\right|$	*d*
2	0.111	0.129	0.018		0.018
5	0.412	0.314	0.098	**0.283**	0.283
6	0.511	0.500	0.011	0.197	0.197
8	0.682	0.686	0.004	0.182	0.182
10	0.809	0.871	0.062	0.123	0.123

Note: Maximum *d* = 0.283.

[a] $F(t_i) = 1 - \exp\left(-\left(\dfrac{t_i}{7.36}\right)^{1.64}\right)$

[b] $\hat{F}(t_i)$ is obtained from the median rank table for *n* = 5.

A2.7   ANALYZING FAILURE DATA WITH SUSPENSIONS

In practice, not every item that is observed is tested to failure. The test may be terminated for reasons other than failure, such as breakdown of the test rig, losing track of a test item, quarantining of the test site or reaching the predetermined time limit for the test.

When we have only partial information about an item's lifetime, the information is known as censored or suspended data. Suspended data will not cause complications in Weibull analysis if all of them are longer than the observed failure times. However, we need to use a special procedure to handle them when some of the failure times are longer than one or more of the suspension times. In the latter case, suspended data are handled by assigning an average order number to each failure time.

Suppose we tested four items, with results shown in Table A2.7. The information shows that the first failure happened at 43 hours. At 55 hours, an item was removed from the test before its failure (it was suspended). Two more failures occurred at 74 and 98 hours, respectively.

If the suspended item had continued to failure, one of the following outcomes shown in Table A2.8 would have resulted. It is noted that the suspended item could have failed in any one of the positions indicated in Table A2.8, which will produce a particular ordering of the failure times. The average position or order number will be assigned to each failure time for plotting.

TABLE A2.7
Data with a Suspended Item

Failure or Suspension	Hours on Test
Failure (F_1)	43
Suspension (S_1)	55
Failure (F_2)	74
Failure (F_3)	98

TABLE A2.8
Three Scenarios for Failure of the Suspended Item

Case I	Case II	Case III
F_1	F_1	F_1
$S_1 \rightarrow F$	F_2	F_2
F_2	$S_1 \rightarrow F$	F_3
F_3	F_3	$S_1 \rightarrow F$

In this example, the first observed failure time will always be in the first position. Thus, it has the order number $m = 1$. As for the second failure time, there are two possibilities that it could have an order number $m = 2$, and one possibility that it would have an order number $m = 3$. Thus, the average order number is

$$m = \frac{3 + 2 \times 2}{3} = 2.33$$

The mean order number of the third failure determined by similar analysis is found to be 3.67.

We can use these mean order numbers to calculate the median rank. For example, using Benard's approximation the median rank for the second failure time (F_2) would be

$$\frac{2.33 - 0.3}{4 + 0.4} = 0.461$$

Alternatively, the median rank for a noninteger order number can be obtained from the median rank table for sample size $n = 4$ by interpolation.

For this example, the data for plotting on Weibull paper are shown in Table A2.9; obviously, using the above procedure to find the mean positions of failure times when there are multiple suspensions is cumbersome. Fortunately, the mean order numbers of failure times intermixed with suspension can be determined by use of the following formula:

$$m_i = m_{i-1} + \frac{(n + 1 - m_{i-1})}{1 + k_i} \tag{A2.1}$$

where m_i = current mean order number, m_{i-1} = previous mean order number, n = total sample size for failure and censorings, and k_i = number of survivors prior to the failure or suspension under consideration.

The following example illustrates the use of Equation A2.1.

TABLE A2.9
Data with Suspended Item

Hours to Failure	Mean Order	Median Rank
43	1.00	0.159
74	2.33	0.461
98	3.67	0.766

EXAMPLE

In a life test program, these failure times were recorded: 31, 39, 57, 65, 70, 105, and 110 hours.

The other items in the sample were removed at the following times without failure (suspension times): 64, 75, 76, 87, 88, 84, 101, 109, and 130 hours.

Even though this sample has 16 items, only 7 failures were observed. To prepare for a Weibull analysis, the data set is reorganized as shown in Table A2.10.

The mean order numbers for the fourth to seventh failures have to be determined.

Consider the fourth failure (F_4), number of survivors prior to the event, $k_4 = 12$. Applying Equation A2.1 gives the mean order number for F_4 as

$$m_4 = m_3 + \frac{(n+1-m_3)}{1+k_4} = 3 + \frac{(16+1-3)}{1+12} = 3 + 1.08 = 4.08$$

Since there is no suspension between F_4 and F_5, the same increment of 1.08 can be applied to determine m_5. Thus, $m_5 = m_4 + 1.08 = 4.08 + 1.08 = 5.16$.

The mean order numbers for F_6 and F_7 are calculated in a similar manner. The data for the Weibull analysis are given in Table A2.11.

TABLE A2.10
Test Data with Multiple Suspensions

	Failure Time, t_i	Suspension Time, s_i	Mean Order Number, m_i
F_1	31		1
F_2	39		2
F_3	57		3
		64	
F_4	65		?
F_5	70		?
		75	
		76	
		84	
		87	
		88	
		101	
F_6	105		?
		109	
F_7	110		?
		130	

TABLE A2.11
Test Data with Multiple Suspensions:
Determining the Median, 5%, and 95% Ranks

	Failure Time, t_i	Suspension Time, s_i	Number of Survivors, k_i	Mean Order Number, m_i	Median Rank (%)	Benard's Approximate (%)	5% Rank	95% Rank
F_1	31		16	1	4.20	4.27	0.30	17.00
F_2	39		15	2	10.20	10.37	2.20	26.00
F_3	57		14	3	16.30	16.46	5.30	34.00
		64	(13)					
F_4	65		12	4.08	22.89	23.05	9.32	42.48
F_5	70		11	5.16	29.49	29.63	13.80	49.12
		75	(10)					
		76	(9)					
		84	(8)					
		87	(7)					
		88	(6)					
		102	(5)					
F_6	105		4	7.53	44.03	44.09	25.65	64.18
		109	(3)					
F_7	110		2	10.69	63.31	63.35	43.14	80.45
		130	(1)					

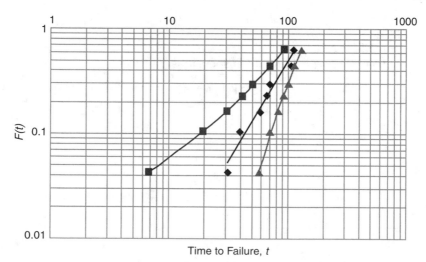

FIGURE A2.18 Weibull plot of the failure data given in Table A2.11.

The median ranks as well as the confidence limits are obtained from the median rank, 5% rank, and 95% rank tables (Appendices 8, 9, and 10, respectively) by interpolation, with sample size $n = 16$. The Weibull plot of the data is shown in Figure A2.18.

A2.8   ANALYZING GROUPED FAILURE DATA WITH MULTIPLE SUSPENSIONS

When there is abundant failure data on an item, we can group the information into separate classes to ease processing.

Suppose that each class interval is of length W. Within an individual class, it could have many failures and suspensions (see Figure A2.19).

We can estimate the hazard rate at the center of the class to be

$$h(t) = \frac{F}{A_V W}$$

where A_V is the average number of items we had in the class interval, that is,

$$A_V = \frac{A + (A - F - C)}{2}$$

When the average hazard rate for each class interval is known, the cumulative distribution function can be estimated by the formula $F(t) = 1 - \exp(-H(t))$, where $H(t)$ is the cumulative hazard function:

$$H(t) = \int_0^t h(t)dt$$

Since we are working with grouped data, the cumulative hazard function takes the form

$$H(t) = \sum h(t) \cdot W$$

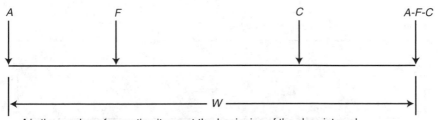

A is the number of operating items at the beginning of the class interval
F is the number of failures in the class interval
C is the number of suspensions in the class interval

FIGURE A2.19 Data in a class interval.

We will study the failure data of a sugar feeder given in the first four columns of Table A2.12. The other columns of the table are calculations for estimating $F(t)$ for individual class intervals.

Figure A2.20 illustrates the calculation for the first class interval (refer to data shown in Table A2.12).

Now that we have the values of $F(t)$ for the end of each class interval, we can plot the data on Weibull paper. The value of $F(t)$ must be plotted at the time corresponding to the end of the interval because it is the cumulative value for the whole interval. To plot $F(t)$ at the mid-point of the interval is wrong because it underestimates the Weibull distribution.

Figure A2.21 shows the Weibull plot of the sugar feeder failure data.

TABLE A2.12
Sugar Feeder Failure Data with Multiple Suspensions

Class, W Weeks	F	C	A	A_v	$h(t)W$	$H(t) = \Sigma h(t)W$	$F(t) = 1 - \exp(-H(t))$
$0 < 1$	9	5	89	82.0	0.110	0.110	0.104
$1 < 2$	16	1	75	66.5	0.241	0.350	0.296
$2 < 3$	9	2	58	52.5	0.171	0.522	0.407
$3 < 4$	7	2	47	42.5	0.165	0.686	0.497
$4 < 5$	2	5	38	34.5	0.058	0.744	0.525
$5 < 6$	2	12	31	24.0	0.083	0.828	0.563
$6 < 7$	3	0	17	15.5	0.194	1.021	0.640
$7 < 8$	2	1	14	12.5	0.160	1.181	0.693
$8 < 9$	2	0	11	10.0	0.200	1.381	0.749
$9 < 10$	0	2	9	8.0			
$10 < 11$	0	0	7	7.0			
$11 < 12$	1	1	7	6.0	0.167	1.548	0.787
$12 < 13$	0	0	5	5.0			
$13 < 14$	1	1	5	4.0	0.250	1.798	0.834
$14 < 15$	1	2	3	1.5	0.667	2.465	0.915

Subtotal   55   34

Total number of items observed = 55 + 34 = 89.

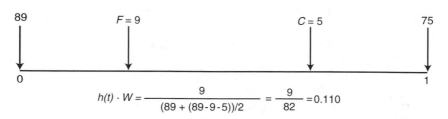

$$h(t) \cdot W = \frac{9}{(89 + (89-9-5))/2} = \frac{9}{82} = 0.110$$

FIGURE A2.20 Data in the first class interval.

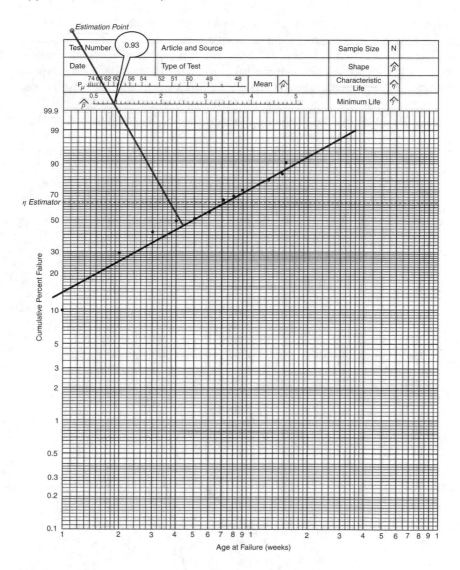

FIGURE A2.21 Weibull plot of the sugar feeder failure data.

A2.9 ANALYZING COMPETING FAILURE DATA

When equipment can fail in more than one way, each of these ways of failure is known as a failure mode. For example, random voltage spikes, which cause failure by overloading the windings, and wear-out failure of bearings are two different failure modes of an electric motor. Since the multiple failure modes are competing with each other to be the event that causes equipment failure, they are known as competing failure modes of the equipment. The failure time distribution of one failure mode may differ from that of another failure mode. Thus, we need to analyze the failure data one failure mode at a time.

Consider a case when there are two failure modes A and B. The time-to-failure data for each failure mode and suspension data are available. We have to apply the following data analysis procedure for determining the failure distribution:

- Do a Weibull plot for failure mode A, treating failures due to failure mode B as suspensions.
- Superimpose a second Weibull plot for failure mode B, treating failures due to failure mode A as suspensions.
- Let $F_A(t)$ and $F_B(t)$ be the cumulative distribution function for modes A and B, respectively; the cumulative distribution function of equipment failure is

$$F_A(t) + F_B(t) - F_A(t) \times F_B(t)$$

This can be derived from the fact that neither mode A nor mode B failure has occurred by time t if equipment is reliable at t.

See Figure A2.22 for an illustration of the Weibull plots.

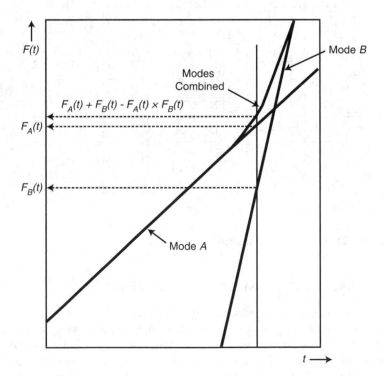

FIGURE A2.22 Weibull plots of competing failure mode data.

A2.10   HAZARD PLOT

For a Weibull distribution with $\gamma = 0$, the hazard rate (also known as instantaneous failure rate) is

$$h(t) = \left(\frac{\beta}{\eta}\right)\left(\frac{t}{\eta}\right)^{\beta-1}$$

Thus, the cumulative hazard function is $H(t) = \int_0^t h(t)dt = \left(\frac{t}{\eta}\right)^{\beta}$

$$\therefore \quad \ln(t) = \frac{\ln\left[H(t)\right]}{\beta} + \ln(\eta)$$

This relationship provides the basis for construction of the Weibull hazard paper, which is basically a log-log paper. Figure A2.23 shows a two-cycle hazard paper for Weibull distributions. The slope of the plot would be $1/\beta$. When $H(t) = 100\%$, $t = \eta$.

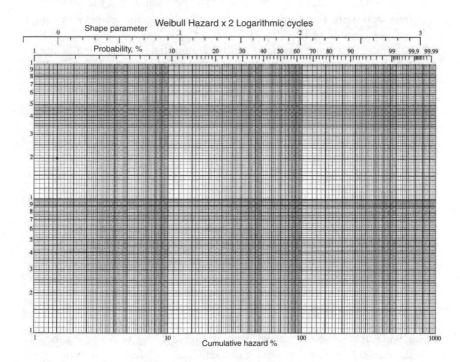

FIGURE A2.23 Two-cycle hazard paper for Weibull distributions.

Instead of plotting the cumulative proportion of items failed, as in the Weibull plot, we now plot the cumulative hazard function, using hazard papers. The plotting procedure is as follows:

1. Tabulate the times to failure in ascending order.
2. For each failure time, t_i, calculate its hazard interval:

$$\Delta H(t_i) = 1 \div (\text{number of items remaining} \\ \text{after the previous failure or censoring})$$

3. For each failure time, calculate the cumulative hazard function:

$$H(t_i) = \Delta H(t_1) + \Delta H(t_2) + \dots + \Delta H(t_{i-1}) + \Delta H(t_i)$$

4. Plot the cumulative hazard against failure time on the chosen hazard paper. If we can fit a straight line through the hazard plot, the Weibull distribution with $\gamma = 0$ can be used as a model of the data set. We can then proceed to estimate the other parameters of the distribution from the plot.
5. From the estimation point located at the intersection of 1.5% cumulative hazard on the X-axis and the value of 20 time units on the Y-axis, draw a line parallel to the fitted line. The value at which the fitted line intersects with the shape parameter scale above the graph gives the estimated value of β.
6. The value of t that corresponds to 100% cumulative hazard on the fitted line is an estimate of η.

EXAMPLE

To construct the hazard plot for the data set given in the example in Table A2.11, we prepare the data shown in Table A2.13.

Figure A2.24 is the hazard plot of this data set. The parameters of the fitted Weibull distribution are $\gamma = 0$, $\beta = 2.09$, and $\eta = 108.8$ hours.

Obviously, the hazard plotting technique has particular advantages when dealing with censored or multifailure mode data. For example, in the latter case, one tabulation may then be used rather than a separate tabulation for each failure mode.

A limitation of hazard plotting is that we cannot construct a confidence interval of the plot.

A2.10.1 NONLINEAR PLOT

As with Weibull probability paper, Weibull hazard paper is based on the two-parameter Weibull distribution, and a nonzero value of failure-free life, γ, will result in a curved cumulative hazard line. A value for γ can be derived by following

TABLE A2.13
Data for a Hazard Plot

Item Number	Number of Survivors	Failure Time	Failure/ Suspension	Δ Hazard (%)	Cumulative Hazard (%)
1	16	31	F	6.25	6.25
2	15	39	F	6.67	12.92
3	14	57	F	7.14	20.06
4	13	64	S		
5	12	65	F	8.33	28.39
6	11	70	F	9.09	37.48
7	10	75	S		
8	9	76	S		
9	8	84	S		
10	7	87	S		
11	6	88	S		
12	5	102	S		
13	4	105	F	25.00	62.48
14	3	109	S		
15	2	110	F	50.00	112.48
16	1	130	S		

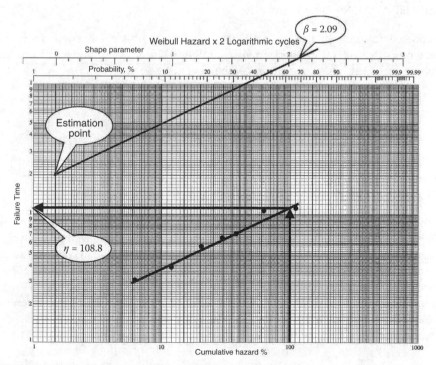

FIGURE A2.24 Hazard plot of data shown in Table A2.13.

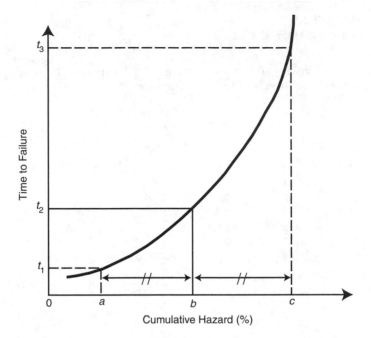

FIGURE A2.25 Nonlinear hazard plot.

a procedure similar to that for the Weibull plot, except that equal length divisions are measured off on the horizontal (cumulative hazard) axis instead of on the vertical axis, as in the case of the Weibull plot (see Figure A2.25).

γ will be estimated by the following expression:

$$\hat{\gamma} = t_2 - \frac{(t_3 - t_2)(t_2 - t_1)}{(t_3 - t_2) - (t_2 - t_1)}$$

A2.11   OTHER APPROACHES TO WEIBULL ANALYSIS

The techniques for Weibull analysis presented in this appendix are based on the regression approach. There are, however, other approaches for Weibull analysis, such as those based on the maximum likelihood criterion or the maximum model accuracy criterion. For details of the maximum likelihood and maximum model accuracy estimation methods, see Lawless (2003) and Ang and Hastings (1994), respectively. Using different approaches for the analysis will give different Weibull models that fit a given data set. Nevertheless, there will be little difference in the fitted model when the data set being analyzed is large.

A2.12　ANALYZING TRENDS OF FAILURE DATA

A Weibull analysis involves fitting a probability distribution to a set of failure data. It is assumed that the process generating the failure times is stable. That means, statistically speaking, all the failure times observed are independently and identically distributed (iid). In reality, this condition may not apply, as in the case of failure times observed from maintenance records of repairable systems. For example, design modifications and improvements made on the equipment in successive life cycles may have the effect of progressively reducing the frequency of failure. In another scenario, imperfect repair or increasing severity of usage in successive life cycles may produce a trend of increasing frequency of failure. Conducting a Weibull analysis on time-between-failure data of these cases is inappropriate because the failure distribution varies from one life cycle to another. The Laplace trend test described below can be used to detect existence of trends in a data set of successive event times.

Let t_i denote the running time of a repairable item at its ith failure, where $i = 1, \ldots, n$, let $N(t_n)$ be the total number of failures observed to time t_n, and the observation terminates at time T when the item is in the operational state. In other words, the failure times are obtained from a *time-terminated test*. Figure A2.26 shows the notations used.

Using the Laplace trend test to determine if the failure events are iid, the test statistic for time-terminated data is

$$u = \sqrt{12N(t_n)} \left(\frac{\sum_1^n t_i}{T \cdot N(t_n)} - 0.5 \right) \qquad (A2.2)$$

If the failure times are iid, u is normally distributed with mean = 0 and standard deviation = 1.

When u is significantly small (negative), we reject the null hypothesis of iid, with the data indicating that there is reliability growth. When u is significantly large (positive), we reject the null hypothesis of iid, with the data indicating that there is reliability deterioration.

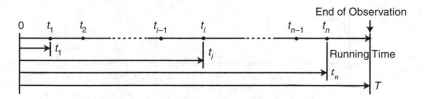

FIGURE A2.26　Time-terminated test data.

If we are satisfied that the failure times are iid, Weibull analysis can be performed on the interfailure times, $(t_i - t_{i-1})$, where $i = 1$ to n.

In the case where the observation terminates at a failure event, say t_n, we have a set of failure-terminated data. The test statistic for failure-terminated data is

$$u = \sqrt{12N(t_{n-1})}\left(\frac{\sum_1^{n-1} t_i}{t_n \cdot N(t_{n-1})} - 0.5 \right) \qquad \text{(A2.3)}$$

EXAMPLE

Machine H fails at the following running times (hours): 15, 42, 74, 117, 168, 233, and 410.

Machine S fails at the following running times (hours): 177, 242, 293, 336, 368, 395, and 410.

Analyze the data and explain the operating behavior of machines H and S.

A2.12.1 MACHINE H

The running times at failure of machine H are displayed graphically in Figure A2.27.

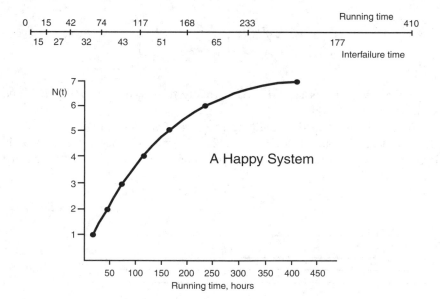

FIGURE A2.27 Failure data of machine H.

This is a set of failure-terminated data. Hence, the test statistic u is calculated using Equation A2.3:

$$\frac{\sum_1^6 t_i}{t_7 \cdot N(t_6)} = \frac{15 + 42 + 74 + 117 + 168 + 233}{410 \times 6} = 0.264$$

$$u = \sqrt{12N(t_6)}\left(\frac{\sum_1^6 t_i}{t_7 \cdot N(t_6)} - 0.5\right) = \sqrt{12 \times 6} \times (0.264 - 0.5) = -2.003$$

At a significance level α of 5%, the lower bound of the test statistic for a two-sided test, $u_{crit.1} = -1.96$. Since u is less than $u_{crit.1}$, we can reject the null hypothesis of iid at $\alpha = 5\%$ and accept the alternate hypothesis that there is reliability growth. Thus, it is not appropriate to perform a Weibull analysis on, or to fit any other probability distribution to, the data set for the purpose of modeling the failure time distribution.

A2.12.2 MACHINE S

The set of interfailure times generated from machine S is identical to that of machine H, except that the sequence is reversed. The running times at failure of machine H are displayed graphically in Figure A2.28.

Using Equation A2.3, we get

$$\frac{\sum_1^6 t_i}{t_7 \cdot N(t_6)} = \frac{177 + 242 + 293 + 336 + 368 + 395}{410 \times 6} = 0.736$$

$$u = \sqrt{12N(t_6)}\left(\frac{\sum_1^6 t_i}{t_7 \cdot N(t_6)} - 0.5\right) = \sqrt{12 \times 6} \times (0.736 - 0.5) = +2.003$$

At a significance level α of 5%, the upper bound of the test statistic for a two-sided test, $u_{crit.2} = +1.96$. Since u is greater than $u_{crit.2}$, we can reject the null hypothesis of iid at $\alpha = 5\%$ and accept the alternate hypothesis that there is reliability deterioration. Thus, it is not appropriate to perform a Weibull analysis on, or to fit any other probability distribution to, the data set for the purpose of modeling the failure time distribution.

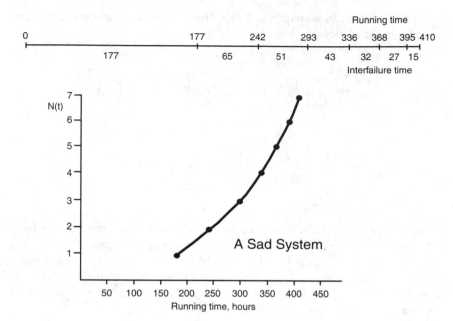

FIGURE A2.28 Failure data of machine S.

REFERENCES

Ang, J.Y.T. and Hastings, N.A.J. (1994). Model accuracy and goodness of fit for the Weibull distribution with suspended items. *Microelectronics and Reliability*, 34, 1177–1184.

Lawless, J.F. (2003). *Statistical Models and Methods for Lifetime Data*, 2nd ed. New York: Wiley.

Murdoch, J. and Barnes, J.A. (1970). *Statistical Tables for Science, Engineering, Management and Business Studies*, 2nd ed. New York: Macmillan.

Nelson, L.S. (1967). Weiboll probability paper, *Journal of Quality Technology*.

O'Connor, P.D. (2002). *Practical Reliability Engineering*, 4th ed. New York: Wiley.

FURTHER READING

Abernethy, R.B. (1996). *The New Weibull Handbook*, 2nd ed. Houston, TX: Gulf Publishing Company.

Ascher, H. and Feingold, H. (1984). *Repairable Systems Reliability*. New York: Marcel Dekker.

Kapur, K.C. and Lamberson, L.R. (1977). *Reliability in Engineering Design*. New York: Wiley.

Murthy, D.N.P., Xie, M., and Jiang, R. (2004). *Weibull Models*. New York: Wiley.

Nelson, W. (2003). *Recurrent Events Data Analysis for Product Repairs, Disease Recurrences, and Other Applications*. Alexandria, VA: American Statistical Association (ASA) and Society for Industrial & Applied Mathematics (SIAM).

Sheskin, D.J. (2004). *Handbook of Parametric and Nonparametric Statistical Procedures*, 3rd ed. London: Chapman & Hall.

Suhir, E. (1997). *Applied Probability for Engineers and Scientists*. New York: McGraw-Hill.

TEAM (Technical and Engineering Aids to Management). (1976). *Technical and Engineering Aids to Management*, Vol. 4, Winter.

PROBLEMS

1. Use Weibull probability paper to determine the parameters of the distribution of the clutch assembly failure data shown in Table A2.14. Also determine the mean life of the assembly.

2. Use Weibull probability paper and the failure data relating to brake pedal bushes shown in Table A2.15 to determine the reliability characteristics of the bushes.

3. The Michyear Tire Company uses 4-cylinder, 4-ton payload delivery vehicles. These vehicles do about 150 miles per day within a 50-mile radius base. There is a suspicion that water pump failures are at an unacceptably high level, and the failure data as shown in Table A2.16 have been obtained.

   Use Weibull probability paper to analyze these data. Using the shape parameter and the characteristic life from your analysis, sketch the

TABLE A2.14
Clutch Assembly Failure Data

Class Interval (in miles, $K = 10^3$)	Number of Failures
0K < 5K	8
5K < 10K	12
10K < 15K	15
15K < 20K	15
20K < 25K	12
25K < 30K	12
30K < 35K	7
35K < 40K	6
40K < 45K	5
45K < 50K	3
50K < 55K	2
55K < 60K	1
60K < 65K	1
65K < 70K	1
Total	100

TABLE A2.15
Brake Pedal Bush Failure Data

Utilization before Failure (in miles, K = 10³)	Frequency of Failures
0K < 5K	0
5K < 10K	2
10K < 15K	2
15K < 20K	4
20K < 25K	4
25K < 30K	5
30K < 35K	6
35K < 40K	6
40K < 45K	7
45K < 50K	5
50K < 55K	7
55K < 60K	8
60K < 65K	4
65K < 70K	6
70K < 75K	4
75K < 80K	6
> 80K	24
Total	100

TABLE A2.16
Water Pump Failure Data

End of Time Interval (in miles, K = 103)	Cumulative Probability $F(t)$ as a %
< 5K	0.00
< 10K	3.08
< 15K	7.96
< 20K	11.40
< 25K	13.19
< 30K	18.67
< 35K	24.21
< 40K	26.13
< 45K	28.33
< 50K	30.00
< 55K	31.20
< 60K	34.92
< 65K	42.00
< 70K	46.00
< 75K	63.25

shape of the associated probability density function, marking on this sketch the mean time to failure of the pumps.

Do you think that preventive replacement of the water pumps may be a worthwhile replacement strategy? Give reasons for your answer.

4. The starter motor failure data shown in Table A2.17 include suspensions. Use Weibull probability paper to determine the reliability characteristics of the motor.

5. A sample of 75 power transistors in germanium are tested and the data collected are given in Table A2.18.

TABLE A2.17
Starter Motor Failure Data

Class Interval (K = 1000 miles)	Number of Failures	Number of Suspensions
0K < 5K	1	0
5K < 10K	1	2
10K < 15K	2	3
15K < 20K	5	2
20K < 25K	1	0
25K < 30K	3	5
30K < 35K	1	1
35K < 40K	1	4
40K < 45K	0	6
45K < 50K	1	2
50K < 55K	1	5
55K < 60K	0	6
60K < 65K	1	6

TABLE A2.18
Power Transistor Test Data

Age to Failure Interval (hours)	Number of Failures
0–250	17
250–500	8
500–750	1
750–1000	1
1000–2000	0
2000–3000	5
3000–4000	3
4000–5000	4
5000–6000	3
6000–7000	2

Graph the failure rate function from $t = 0$ to 10,000 hours.

What kind of failure do you suspect we have? What would you suggest to improve the reliability?

6. The data in Table A2.19 represent the cycles to failure of a small electrical appliance.

TABLE A2.19
Failure Data of an
Electrical Appliance

Time in Cycles	Event
1430	1 censored
1624	1 censored
1877	1 censored
2615	1 censored
3075	1 censored
3174	1 censored
3264	1 censored
3424	1 censored
3508–4161	16 censored
4552–4589	12 censored
1015	1 failure
1493	1 failure
1680	1 failure
2961	1 failure
2974	1 failure
3009	1 failure
3244	1 failure
3462	1 failure
4246	1 failure

---

a. Make a graph of the cumulative failure rate function and estimate the parameters of the Weibull distribution.

b. From the graph, find $R(1000 \text{ cycles}|3000 \text{ cycles})$.

7. A diesel engine was monitored onboard a ship over a period of 10 years, and Figure A2.29 indicates the failure pattern of the engine. It is assumed that after each failure the engine is returned to the as-new condition by maintenance.

At time = 10 years you are asked to analyze the failure statistics and give an estimate of the Weibull parameters β and η. Assume $\gamma = 0$ and comment generally on the engine's performance. Specifically you must answer the following questions:

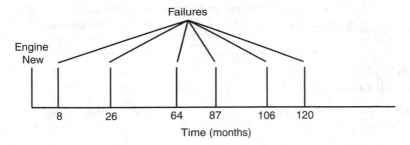

FIGURE A2.29 Engine failure pattern.

    a. Using Weibull probability paper, what are your best estimates of β and η?

    b. Judge whether or not your sample data can be represented by a Weibull distribution by using the Kolmogorov–Smirnov goodness-of-fit test.

Problems 8 to 10 require the use of the WeibullSoft package, which can be downloaded from http://www.crcpress.com/e_products/downloads/download.asp?cat_no=DK9669.

  8. You are given a set of failure data for heavy-duty bearings in a steel forging plant with failure times (in hours) as follows:

TABLE A2.20
Bearing Failure Data

| 2082 | 1717 | 2263 | 3945 | 5093 | 2751 | 3065 | 12,456 | 1340 | 7062 |

Fit a two-parameter Weibull model to the data set.

    a. What are the two parameters used in this analysis, and what are their values? What do these parameters represent?

    b. What are the values for μ and σ?

    c. At 10% significance level, can it be accepted that the chosen distribution fits the data?

    d. What is the probability that the bearings will fail before time 3000?

    e. What is the B_{10} life of these bearings?

Now, fit a three-parameter Weibull model to the data set.

    f. What is the value of the third parameter in this analysis? What does this parameter represent?

    g. What are the values of μ and σ in this distribution?

9. The service life of certain fan belts has been monitored and recorded, with times (in weeks) as follows (F = failure, S = suspension):

TABLE A2.21
Fan Belt Failure Data

| 174(F) | 124(F) | 106(F) | 153(F) | 160(F) | 167(F) | 112(F) | 194(F) | 181(F) | 136(S) |

---

   a. How many data points should be used in the Weibull analysis? Why?
   b. What are the values of d and d_α for a goodness-of-fit test in the Weibull analysis? Will you reject the hypothesis at a 10% significance level that the two-parameter Weibull distribution determined by the package is a model of the data set?
   c. Fit a three-parameter Weibull model to the data set. Give the values of β, η, γ, μ, and σ of the model.

10. You are given the following failure data for the lifetime of a certain lightbulb (in hours):

TABLE A2.22
Light Bulb Failure Data

| 3129 | 1593 | 7427 | 8968 | 4019 | 5188 | 7239 | 3662 | 2876 | 5817 |

---

   a. In the three-parameter analysis, what does $F(t - \gamma)$ represent? What is the value of $F(t - \gamma)$ for $t = 5000$?
   b. What is the point estimate of t when $F(t - \gamma)$ is 20%? Determine the 90% confidence interval of t.

11. A component installed in a photocopying machine experienced frequent failures. Failure records in cumulative copies at which failures of this component occurred are listed below:

TABLE A2.23
Photocopying Failure Data

| 12,204 | 21,384 | 26,909 | 33,912 | 38,232 | Current cumulative copies = 40,500 |

---

   a. Perform an appropriate statistical test to detect the existence of a trend in the failure times of this component. Use a significance level of 5%.

      What is the implication of the findings from the test?

b. Use a graphical method to determine the parameters of the model that fits the failure data. Sketch the probability density function and hazard rate function of the fitted model.

If the component is to be at least 90% reliable at 5000 copies, does it meet this design objective?

12. A bearing may fail in one of two modes: ball failure or inner race failure. Data from a bearing life study program conducted on a sample of 10 units are given in Table A2.24.

TABLE A2.24
Bearing Failure Data

Specimen Number	Hours to Failure	Failure Mode
1	8	Ball
2	50	Ball
3	102	Ball
4	224	Ball
5	22	Ball
6	140	Ball
7	120	Inner race
8	20	Inner race
9	300	Inner race
10	90	Inner race

Suppose you are going to use a graphical method for data analysis. Show how you would process the data to determine the distribution of hours to failure for the two modes of bearing failure.

Use a suitable probability paper to produce a plot of the above data set. Use the information obtained from the plot to estimate the reliability of the bearing after 100 hours of usage. State any assumption(s) used in your analysis.

Appendix 3: Time Value of Money: Discounted Cash Flow Analysis

The interest rate relevant for a firm's decision-making is an important subject in its own right and is a lively topic of concern among scholars and practitioners of finance.

H.M. Wagner

A3.1 INTRODUCTION

The purpose of this appendix is to present key aspects of engineering economics that are relevant to the topic of establishing the economic life of capital equipment that is contained in Chapter 4. See texts by DeGarmo et al. (1986) and Park et al. (2000) for a comprehensive discussion of engineering economics.

The basic problem is illustrated in Figure A3.1. Since the economic life of capital equipment will be measured in years, rather than months, as may be the case for the component replacement problems discussed in Chapter 2, it is necessary to take into account the fact that, in the economic life calculation, money changes in value over time. Also, there are concerns about the effect of inflation and tax issues in the analysis of capital equipment problems. These matters will be addressed in this appendix.

Many maintenance decisions, such as that to replace an expensive piece of plant, involve the investment of large sums of money. The costs and benefits accruing from the investment will continue for a number of years. When the investment of money today influences cash flows in the future, and when we are evaluating alternative investment opportunities, account needs to be taken of the fact that the value now of an amount of money depends on when that amount is due to be paid or received. For example, $100 received in the future is worth less than $100 received now. To enable comparisons of alternative investments to be made, we convert the value of money that is either to be spent or to be received in the future as a result of the investment, into its present-day value; we determine the present value (or present worth) of the investment decision. The present value criterion summarizes in one numerical index the value of a stream of cash flows even if we consider an infinite series of cash flows, and so enables alternative investments to be ranked in order of preference; although with some investment decisions, the uncertainties of future events are so great that any sophisticated analysis is not worthwhile.

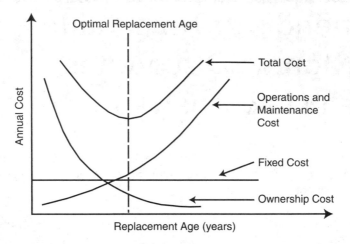

FIGURE A3.1 The economic life problem.

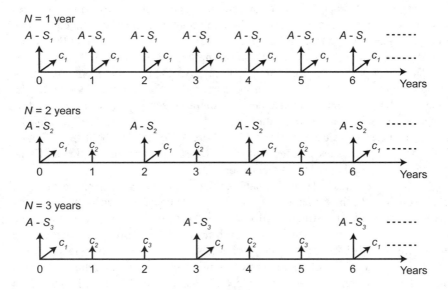

FIGURE A3.2 Concept of an asset's economic life.

For example, consider Figure A3.2. If an asset is replaced every year ($N = 1$), then there is a certain cash flow as depicted in the top figure — A is the purchase price, C_i is the operations and maintenance costs in the ith year of the asset's life, and S_i is the resale (or scrap) value of an asset of age i. But if the asset is replaced on a 2- or 3-year cycle, the cash flow shown in the middle and bottom figures in Figure A3.2 will be different. Clearly, it is necessary to compare all the possibilities fairly, and it is for this reason that we often evaluate alternative replacement cycles on the basis of their present value (sometimes called total discounted cost or net present value). In addition to calculating the present values associated with alternative streams of cash flow, we also need to consider them over the same planning horizon. Alternatively, their equivalent annual cost (EAC) needs to be calculated (see Section A3.5).

A3.2   PRESENT VALUE FORMULAS

To introduce the present value criterion, consider the following. If a sum of money, say $1000, is deposited in a bank where the compound interest rate on such deposit is 10% per annum, payable annually, then after 1 year there will be $1100 in the account. If this $1100 is left in the account for a further year, there will then be $1210 in the account.

In symbolic notation we are saying that if $L is invested and the relevant interest rate is $i\%$ per annum, payable annually, then after n years the sum S resulting from the initial investment is

$$S = \$L\left(1 + \frac{i}{100}\right)^n \qquad (A3.1)$$

Thus, if $L = 1000$, $i = 10\%$, and $n = 2$ years,

$$S = 1000(1 + 0.1)^2 = \$1210$$

The present-day value of a sum of money to be spent or received in the future is obtained by doing the reverse calculation. Namely, if $\$S$ is to be spent or received n years in the future, and $i\%$ is the relevant interest rate, then the present value of $\$S$ is

$$PV = \$S\left(\frac{1}{1 + i/100}\right)^n \qquad (A3.2)$$

where $\left(\dfrac{1}{1 + i/100}\right)^n = r$ is termed the discount factor.

Thus, the present-day value of $1210 to be received 2 years from now is

$$PV = 1210\left(\frac{1}{1 + 0.1}\right)^2 = \$1000$$

that is, $1000 today is equivalent to $1210 2 years from now when $i = 10\%$ per annum.

It has been assumed that the interest rate is paid once per year. In fact, the interest rate may be paid weekly, monthly, quarterly, semiannually, and so on, and when this is the case, Equations A3.1 and A3.2 are modified as follows: If the nominal interest rate* is $i\%$ per annum, payable m times per year, then in n years the value $$S$ of an initial investment of $$L$ is

$$S = \$L\left(1 + \frac{i/100}{m}\right)^{nm} \tag{A3.3}$$

Thus, the present value of $$S$ to be spent or received n years in the future is

$$PV = \$S\left(\cfrac{1}{1 + \cfrac{i/100}{m}}\right)^{nm} \tag{A3.4}$$

It is also possible to assume that the interest rate is paid continuously. This is equivalent to letting m in Equation A3.3 tend to infinity. When this is the case,

$$\lim_{m \to \infty} L\left(1 + \frac{i/100}{m}\right)^{nm} = \$L\exp\left[\frac{in}{100}\right] \tag{A3.5}$$

and the appropriate present value formula is

$$PV = \$S\exp\left[-\frac{in}{100}\right] \tag{A3.6}$$

In practice, with capital equipment replacement problems, it is customary to assume that interest rates are payable once per year, and so Equation A3.2 is used in present value calculations. Continuous discounting is sometimes used for its mathematical convenience, or because it is thought that it reflects cash flows more accurately. If this is the case, then Equation A3.6 is used.

It is customary to assume that the interest rate i is given as a decimal, and not in percentage terms. Equations A3.2 and A3.6 are then written as

---

* Sometimes interest is compounded at time intervals shorter than 1 year. However, interest rates are typically stated on an annual basis. For example, the interest rate can be 2.5% compounded quarterly. In such case, the *nominal annual rate of interest* is 10%. It should be noted that in this example, the *actual annual rate of interest* is greater than 10%.

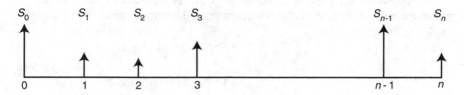

FIGURE A3.3 Cash flows.

$$PV = \$S\left(\frac{1}{1+i}\right)^n \qquad (A3.7)$$

$$PV = \$S \exp[-in] \qquad (A3.8)$$

Both of the above formulas are used in some of the replacement problems discussed in Chapter 4.

An illustration of the sort of problems where the present value criterion is used is the following (Figure A3.3): if a series of payments S_0, S_1, S_2, ..., S_n, illustrated in Figure A3.1, are to be made annually over a period of n years, then the present value of such a series is

$$PV = S_0 + S_1\left(\frac{1}{1+i}\right)^1 + S_2\left(\frac{1}{1+i}\right)^2 + ... + S_n\left(\frac{1}{1+i}\right)^n \qquad (A3.9)$$

If the payments S_j, where $j = 0, 1, ..., n$, are equal, then the series of payments is termed an annuity and Equation A3.9 becomes

$$PV = S + S\left(\frac{1}{1+i}\right) + S\left(\frac{1}{1+i}\right)^2 + ... + S\left(\frac{1}{1+i}\right)^n \qquad (A3.10)$$

which is a geometric progression, and the sum of $n + 1$ terms of a geometric progression gives

$$PV = S\left[\frac{1-\left(\frac{1}{1+i}\right)^{n+1}}{1-\left(\frac{1}{1+i}\right)}\right] = S\left(\frac{1-r^{n+1}}{1-r}\right) \qquad (A3.11)$$

If the series of payments of Equation A3.10 is assumed to continue over an infinite period of time, that is, $n \to \infty$, then from the sum to infinity of a geometric progression we get

$$PV = \frac{S}{1-r} \tag{A3.12}$$

Using continuous discounting, the equivalent expression to Equation A3.9 is

$$PV = S_0 + S_1 \exp[-i] + S_2 \exp[-2i] + \ldots + S_n \exp[-ni] \tag{A3.13}$$

Again, if the S values are equal, we have the sum of $n + 1$ terms of a geometric progression, which gives

$$PV = S\left(\frac{1 - \exp[-(n+1)i]}{1 - \exp[-i]}\right) \tag{A3.14}$$

If the series of payments is assumed to continue over an infinite period, we get

$$PV = \frac{S}{1 - \exp[-i]} \tag{A3.15}$$

In all of the above formulas we have assumed that i remains constant over time. If this is not a reasonable assumption, then Equations A3.9 and A3.13 need to be modified slightly; for example, we may let i take the values i_1, i_2, ..., in the different periods.

A3.3   DETERMINATION OF APPROPRIATE INTEREST RATE

In practice, it is necessary to know the appropriate value of interest rate i to use in any present value calculation. Often difficulties are encountered when attempting to specify this value. If money is borrowed to finance the investment, then the value of i used in the calculations is the interest rate paid on the borrowed money. If the investment is financed by the internal resources of a company, then i is related to the interest rate obtained from investments within the company.

In a survey in the United States of companies regarding the way in which the interest rate, also known as the discount rate, is calculated, the following was stated: "31% of the firms used the rate of return on new investments. 26% used the weighted average of market yields on debt and equity securities. . . . 18% of the firms used the cost of additional borrowing. . . . 6% used the rate which keeps the market price of a common stock of the firm from falling."

As Wagner (1969) says: "The interest rate relevant for a firm's decision-making is an important subject in its own right and is a lively topic of concern among scholars and practitioners of finance."

As far as the present value criterion is concerned, we will assume that an appropriate value of i can be specified. Difficulties associated with uncertainty in i can often be reduced by the use of sensitivity analysis, and some comments on this are made in Section 2.2.4.

A3.4   INFLATION

From an examination of Figure A3.2 it can be seen that the acquisition cost of the asset is denoted by the symbol A. Consider the 3-year replacement cycle: in this case it is suggested that the purchase price remains the same in years 3 and 6 as it was in year 1. Mathematically, A cannot change its value. If inflation is taking place, this is clearly not true. The price in year 3 will be the price in year 1 plus the effect of inflation. However, it can be shown that, provided inflation is occurring at a constant rate, the present value of a future stream of cash flows is the same regardless of whether the effect of inflation is built into the future cash flows. However, if nominal dollars are used (dollars having the value of the year in which they are spent or received), then the interest rate used for discounting purposes must take inflation into account and build the effect of inflationary factors into future cash flow estimates.

In practice, most organizations undertake their capital equipment replacement analyses using real dollars (dollars having present-day value) and use an inflation-free or real interest rate for discounting purposes.

A3.5   THE EQUIVALENT ANNUAL COST

Equation A3.9 states that if the payments S_i were equal, we would have an annuity. When calculating the present values associated with a stream of cash flow associated with purchasing, operating, maintaining, and eventually disposing of an asset — namely, the life cycle costs — there are peaks and troughs in the cash flows. They certainly are not equal each year. The present value calculation brings all these future cash flows to a single number, the present value (PV). For management decision making, it is usually more meaningful to present that PV in terms of its EAC, which can be thought of as the annuity value. In other words, the EAC smoothes out the peaks and troughs in the various cash flows and converts them to an equivalent equal cash flow for each year; it might be thought of as the amount of funds an organization is required to put into its budget each year to fund the purchase, operation, maintenance, and disposal of an asset according to a specified asset replacement policy. The EAC is discussed in Chapter 4. To convert the PV to its EAC, the PV is multiplied by the capital recovery factor (CRF):

TABLE A3.1
Machine Tool Cash Flow

Machine Tool	Purchase Price ($)	Installation Cost ($)	Operating Cost ($)			Salvage Value
			Year 1	Year 2	Year 3	
A	5000	100	100	100	100	3000
B	3000	100	200	300	400	1500
C	6000	100	50	80	100	3500

Note: Cost in $ × 100.

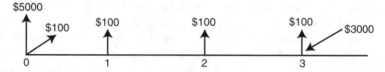

FIGURE A3.4 Cash flow for machine tool A.

$$CRF = \frac{i(1+i)^n}{(1+i)^n - 1} \qquad (A3.16)$$

where i is the interest rate appropriate for discounting (real or interest free) and n is the number of years over which the discounting occurs. This will be illustrated in the following example.

A3.6   EXAMPLE: SELECTING AN ALTERNATIVE — A ONE-SHOT DECISION

To illustrate the application of the present value criterion and the equivalent annual cost when deciding which is the best from a set of possible investment opportunities, we will consider the following problem.

A subcontractor obtains a contract to maintain specialized equipment for a period of 3 years, with no possibility of an extension of this period. To cope with the work, the contractor has to purchase a special-purpose machine tool. Given the costs and salvage values shown in Table A3.1 for three equally effective machine tools, which one should the contractor purchase? We will assume that the interest rate appropriate for discounting is 11% and that operating costs are paid at the end of the year in which they are incurred.

For machine tool A the cash flow is depicted in Figure A3.4.

$$\text{Present value} = \$5000 + \$100 + \$100(0.9)$$
$$+ \$100(0.9)^2 + \$100(0.9)^3 - \$3000(0.9)^3 = \$3157$$

Recall the discount factor, $r = 1/(1 + i)$. Since $i = 11\%$, then r (as a decimal fraction) = 0.9.

Similarly, for machine tool B,

$$\text{Present value} = \$3000 + \$100 + \$200(0.9)$$
$$+ \$300(0.9)^2 + \$400(0.9)^3 - \$1500(0.9)^3 = \$2721$$

and for machine tool C,

$$\text{Present value} = \$6000 + \$100 + \$50(0.9)$$
$$+ \$80(0.9)^2 + \$100(0.9)^3 - \$3500(0.9)^3 = \$3731$$

Thus, equipment B should be purchased since it gives the minimum present value of the costs, namely, \$2721.

Note: If the time value of money had not been taken into account in the evaluation of the three choices, the costs would be given as follows:

Machine tool A: \$2400
Machine tool B: \$2500
Machine tool C: \$2830

And so machine tool A would have been selected as the best buy. In practice, organizations evaluate alternatives through taking into account the time value of money, such as by calculating the PV associated with the various alternatives.

Rather than evaluating the alternatives by providing the PV associated with each of them, the EAC could have been calculated. Recall from Section A3.5 that

$$\text{EAC} = \text{PV} \times \text{CRF}$$

where $CRF = \dfrac{i(1+i)^n}{\left(1+i\right)^n - 1}$.

For machine tool A, we had a PV = \$3157.

$$EAC = 3157\left[\frac{i(1+i)^n}{\left(1+i\right)^n - 1}\right]$$

$$= 3157\left[\frac{0.11\left(1+0.11\right)^3}{\left(1+0.11\right)^3 - 1}\right]$$

$$= 3157 \times 0.4092$$

$$= \$1291.89$$

Graphically, we have

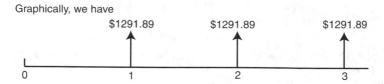

FIGURE A3.5 Machine tool A's EAC.

This is illustrated graphically in Figure A3.5. The present value of the annuity of $1291.89 for the 3 years is the same as the original stream of cash flows.

A3.7 FURTHER COMMENTS

In the above machine tool purchasing example it will be noticed that the same decision on the tool to purchase would not have been reached if no account had been taken of the time value of money. Note also that many of the figures used in such an analysis will be estimates of future costs or returns. Where there is uncertainty about any such estimates, or where the PV calculation indicates several equally acceptable choices (because their PVs are more or less the same), a sensitivity analysis of some of the estimates may provide information to enable an obvious decision to be made. If this is not possible, then we may invoke other factors, such as familiarity with supplier, availability of spares, and so on, to assist in making the decision. Of course, when estimating future costs and returns, account should be taken of possible increases in materials costs, wages, and the like (i.e., inflationary factors).

When dealing with capital investment decisions, a criterion other than PV is sometimes used. For discussion of such criteria, for example, payback period and internal rate of return, the reader is referred to the engineering economics literature, such as Park et al. (2000) and DeGarmo et al. (1986).

REFERENCES

DeGarmo, E.P., Sullivan, W.G., and Canada, J.G. (1986). *Engineering Economy*, 7th ed. New York: Macmillan.
Park, C., Pelot, R., Porteous, K., and Zuo, M.J. (2000). *Contemporary Engineering Economics: A Canadian Perspective*, 2nd ed. Pearson Education.
Wagner, H.M. (1969). *Principles of Operations Research*. Englewood Cliffs, NJ: Prentice Hall.

Appendix 4: List of Applications of Maintenance Decision Optimization Models

> The proof of the pudding is in the eating.
>
> **An English idiom**

Within each of Chapters 2 to 5, there are sections highlighting applications of the theory contained within the chapter. The following is an expanded list of applications with only titles provided. The purpose of the list, which is not exhaustive, is to illustrate the breadth of applications with which the authors have been associated that used the models presented in this book, or their extensions. The authors have been directly involved in each study, either as an advisor to an organization or through supervising undergraduate or postgraduate students as they undertook a project as part of their studies. A number of the projects were undertaken by postexperience students who have taken courses from the authors as part of company training programs in the general area of maintenance optimization and reliability engineering (MORE). As part of the training program, course participants worked in teams and undertook pilot studies applying the ideas contained in this book.

Aluminum and steelmaking
- Optimization of nitrogen compressor third-stage piston ring: condition-based maintenance (CBM) model
- Establishing the economic life of mobile equipment (floor sweepers, forklift trucks, and General Motors Suburban)
- Establishing lathe requirements in a steel mill

Electrical generation
- Reactor coolant pump: CBM model
- Optimizing CBM decisions on main rotating equipment using proportional hazard model
- Air preheater cleaning
- Condition monitoring of hydrodyne seals in a nuclear plant
- Maximizing power station reliability subject to a budgetary constraint

Electricity transmission and distribution
- Optimal preventive replacement of electronic modules in 110-V DC battery chargers
- Optimal preventive replacement of unloader units in air compressors
- Optimal preventive replacement of capacitor units in a capacitor bank
- Replacement of 400-kV lightning arrestors
- Rightsizing of cable joining resources to meet a fluctuating workload, taking into account the subcontracting opportunities
- Optimizing inspection frequency for an overhead line supervisory information system
- Optimizing inspection frequency for air compressor systems
- Determining the number of cable oil vans to meet service demand
- BP2 deionized water pump replacement
- Serviceability life expectancy study of built-up roofs
- HVDC valve damping equipment failures
- Optimal replacement age of fast gas relays
- Replacement of fault detector relays
- Transformer repair vs. replace decision analysis
- Optimal number of spare transformers
- Economic life of 230-kV oil circuit breakers

Food-processing industries
- Sugar refinery centrifuge component replacement
- Condition monitoring of shear pump bearings in a canning plant
- Establishing the economic refurbishment time for a seamer in a canning plant

Military (land, sea, and air)
- Oil analysis of marine diesel engines: optimizing condition-based maintenance decisions (U.K.)
- Optimizing condition-based maintenance decisions: an application to ship diesel engines (Canada)
- Aircraft fuel pump replacement policy
- Condition monitoring of aircraft engines subject to oil analysis

Mining industry
- Optimal inspection frequency for scissor lift vehicles
- Components for preventive replacement of McLean bolters
- Optimization of (100t loco) inspection frequency
- Economic life analysis of loader
- Optimizing availability of mill GIW discharge pumps
- Optimizing availability of smelter converters
- Establishing the economic life of a fleet of haul trucks
- Steering clutch replacement of a dozer
- Spares provisioning of electric motors on conveyor systems
- Condition monitoring of engines and transmissions on haul trucks
- Condition monitoring of electric wheel motors

- Condition monitoring of pump bearings in a coal plant
- Shovel replacement in light of technological improvement
- Repair vs. replacement for a wheel loader
- Optimizing the number of vehicles in a haul truck fleet

Oil and gas

- Economic life of a combustion engine
- Cylinder head replacement
- Compressor valve replacement
- Pressure safety valve inspection interval
- Optimizing number of spare 100-hp motors
- CBM optimization of engine pumping unit
- Maintenance crew optimization
- Condition monitoring of oil well pumping system (casing, sucker rod, and pump)

Petrochemical industry

- Optimizing maintenance crew size and shift patterns

Pharmaceutical industry

- Failure-finding interval for a compressor: parallel redundant system
- Huber washer replacement policy
- Work center resource optimization

Pulp and paper industry

- Recovery soot blower component replacement strategy: lance tube packing failures
- Bark hog equipment failure analysis: establishing productive maintenance (PM) interval
- Tissue machine tail cutter: drive belt replacement policy
- Sawmill sawquip line: replacement policy for outfeed press roll bearings
- Establishing the economic life of a feller-buncher

Railway systems

- Optimizing condition-based maintenance decisions to reduce in-service failures of traction motor ball bearings

Transportation

- Forklift truck replacement cycles
- Transit bus fleet replacement policy
- Establishing the economic life and optimal maintenance policy for a fleet of trailers
- Transit bus fleet inspection policy: A, B, C, and D class inspections
- Establishing the economic life of a fleet of tractors

Appendix 5: Ordinates of the Standard Normal Distribution

The table gives $\phi(z)$ for values of the standardized normal variate, z, in the interval

$$\phi(z) = \frac{1}{\sqrt{2\pi}} \exp\left(\frac{-z^2}{2}\right)$$

0.0(0.1)4.0, where

$$z = \frac{t - \mu}{\sigma}$$

for a normal distribution with mean = μ and standard deviation = σ.

z	0.0	0.1	0.2	0.3	0.4	0.5	0.6	0.7	0.8	0.9
0.0	0.3989	0.3970	0.3910	0.3814	0.3683	0.3521	0.3332	0.3123	0.2897	0.2661
1.0	0.2420	0.2179	0.1942	0.1714	0.1497	0.1295	0.1109	0.0940	0.0790	0.0656
2.0	0.0540	0.0440	0.0355	0.0283	0.0224	0.0175	0.0136	0.0104	0.0079	0.0060
3.0	0.0044	0.0033	0.0024	0.0017	0.0012	0.0009	0.0006	0.0004	0.0003	0.0002
4.0	0.0001									

Source: Murdoch, J. and Barnes, J.A., *Statistical Tables for Science, Engineering, Management and Business Studies*, 2nd ed., Macmillan, New York, 1970. (Reproduced with permission of Palgrave Macmillan.)

Appendix 6:
Areas in the Tail of the
Standard Normal Distribution

The function tabulated is $1 - \Phi(z)$, where $\Phi(z)$ is the cumulative distribution function of a standardized normal variate, z. Thus,

$$1 - \Phi(z) = \frac{1}{\sqrt{2\pi}} \int_z^\infty \exp\left(\frac{-x^2}{2}\right) dx$$

is the probability that a standardized normal variate selected at random will be greater than a value of $z\left(> \dfrac{x - \mu}{\sigma} \right)$ (Figure A6.1).

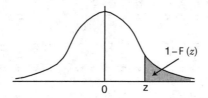

FIGURE A6.1

Areas in the Tail of the Standard Normal Distribution

$\frac{x-\mu}{\sigma}$	0.00	0.01	0.02	0.03	0.04	0.05	0.06	0.07	0.08	0.09
0.0	0.5000	0.4960	0.4920	0.4880	0.4840	0.4801	0.4761	0.4721	0.4681	0.4641
0.1	0.4602	0.4562	0.4522	0.4483	0.4443	0.4404	0.4364	0.4325	0.4286	0.4247
0.2	0.4207	0.4168	0.4129	0.4090	0.4052	0.4013	0.3974	0.3936	0.3897	0.3859
0.3	0.3821	0.3783	0.3745	0.3707	0.3669	0.3632	0.3594	0.3557	0.3520	0.3483
0.4	0.3446	0.3409	0.3372	0.3336	0.3300	0.3264	0.3228	0.3192	0.3156	0.3121
0.5	0.3085	0.3050	0.3015	0.2981	0.2946	0.2912	0.2877	0.2843	0.2810	0.2776
0.6	0.2743	0.2709	0.2676	0.2643	0.2611	0.2578	0.2546	0.2514	0.2483	0.2451
0.7	0.2420	0.2389	0.2358	0.2327	0.2296	0.2266	0.2236	0.2206	0.2177	0.2148
0.8	0.2119	0.2090	0.2061	0.2033	0.2005	0.1977	0.1949	0.1922	0.1894	0.1867
0.9	0.1841	0.1814	0.1788	0.1762	0.1736	0.1711	0.1685	0.1660	0.1635	0.1611
1.0	0.1587	0.1562	0.1539	0.1515	0.1492	0.1469	0.1446	0.1423	0.1401	0.1379
1.1	0.1357	0.1335	0.1314	0.1292	0.1271	0.1251	0.1230	0.1210	0.1190	0.1170
1.2	0.1151	0.1131	0.1112	0.1093	0.1075	0.1056	0.1038	0.1020	0.1003	0.0985
1.3	0.0968	0.0951	0.0934	0.0918	0.0901	0.0885	0.0869	0.0853	0.0838	0.0823
1.4	0.0808	0.0793	0.0778	0.0764	0.0749	0.0735	0.0721	0.0708	0.0694	0.0681
1.5	0.0668	0.0655	0.0643	0.0630	0.0618	0.0606	0.0594	0.0582	0.0571	0.0559
1.6	0.0548	0.0537	0.0526	0.0516	0.0505	0.0495	0.0485	0.0475	0.0465	0.0455
1.7	0.0446	0.0436	0.0427	0.0418	0.0409	0.0401	0.0392	0.0384	0.0375	0.0367
1.8	0.0359	0.0351	0.0344	0.0336	0.0329	0.0322	0.0314	0.0307	0.0301	0.0294
1.9	0.0287	0.0281	0.0274	0.0268	0.0262	0.0256	0.0250	0.0244	0.0239	0.0233

	0.00	0.01	0.02	0.03	0.04	0.05	0.06	0.07	0.08	0.09
2.0	0.02275	0.02222	0.02169	0.02118	0.02068	0.02018	0.01970	0.01923	0.01876	0.01831
2.1	0.01786	0.01743	0.01700	0.01659	0.01618	0.01578	0.01539	0.01500	0.01463	0.01426
2.2	0.01390	0.01355	0.01321	0.01287	0.01255	0.01222	0.01191	0.01160	0.01130	0.01101
2.3	0.01072	0.01044	0.01017	0.00990	0.00964	0.00939	0.00914	0.00889	0.00866	0.00842
2.4	0.00820	0.00798	0.00776	0.00755	0.00734	0.00714	0.00695	0.00676	0.00657	0.00639
2.5	0.00621	0.00604	0.00587	0.00570	0.00554	0.00539	0.00523	0.00508	0.00494	0.00480
2.6	0.00466	0.00453	0.00440	0.00427	0.00415	0.00402	0.00391	0.00379	0.00368	0.00357
2.7	0.00347	0.00336	0.00326	0.00317	0.00307	0.00298	0.00289	0.00280	0.00272	0.00264
2.8	0.00256	0.00248	0.00240	0.00233	0.00226	0.00219	0.00212	0.00205	0.00199	0.00193
2.9	0.00187	0.00181	0.00175	0.00169	0.00164	0.00159	0.00154	0.00149	0.00144	0.00139
3.0	0.00135									
3.1	0.00097									
3.2	0.00069									
3.3	0.00048									
3.4	0.00034									
3.5	0.00023									
3.6	0.00016									
3.7	0.00011									
3.8	0.00007									
3.9	0.00005									
4.0	0.00003									

Source: Murdoch, J. and Barnes, J.A., *Statistical Tables for Science, Engineering, Management and Business Studies*, 2nd ed., Macmillan, New York, 1970. (Reproduced with permission of Palgrave Macmillan.)

Appendix 7:
Values of Gamma Function

n	Γ(n)	n	Γ(n)	n	Γ(n)	n	Γ(n)
1.00	1.00000	1.25	0.90640	1.50	0.88623	1.75	0.91906
1.01	0.99433	1.26	0.90440	1.51	0.88559	1.76	0.92137
1.02	0.98884	1.27	0.90250	1.52	0.88704	1.77	0.92376
1.03	0.98355	1.28	0.90072	1.53	0.88757	1.78	0.92623
1.04	0.97844	1.29	0.89904	1.54	0.88818	1.79	0.92877
1.05	0.97350	1.30	0.89747	1.55	0.88887	1.80	0.93138
1.06	0.96874	1.31	0.89600	1.56	0.88964	1.81	0.93408
1.07	0.96415	1.32	0.89464	1.57	0.89049	1.82	0.93685
1.08	0.95973	1.33	0.89338	1.58	0.89142	1.83	0.93969
1.09	0.95546	1.34	0.89222	1.59	0.89243	1.84	0.94261
1.10	0.95135	1.35	0.89115	1.60	0.89352	1.85	0.94561
1.11	0.94740	1.36	0.89018	1.61	0.89468	1.86	0.94869
1.12	0.94359	1.37	0.88931	1.62	0.89592	1.87	0.95184
1.13	0.93993	1.38	0.88854	1.63	0.89724	1.88	0.95507
1.14	0.93642	1.39	0.88785	1.64	0.89864	1.89	0.95838
1.15	0.93304	1.40	0.88726	1.65	0.90012	1.90	0.96177
1.16	0.92980	1.41	0.88676	1.66	0.90167	1.91	0.96523
1.17	0.92670	1.42	0.88636	1.67	0.90330	1.92	0.96877
1.18	0.92373	1.43	0.88604	1.68	0.90500	1.93	0.97240
1.19	0.92089	1.44	0.88581	1.69	0.90678	1.94	0.97610
1.20	0.91817	1.45	0.88566	1.70	0.90864	1.95	0.97988
1.21	0.91558	1.46	0.88560	1.71	0.91057	1.96	0.98374
1.22	0.91311	1.47	0.88563	1.72	0.91258	1.97	0.98768
1.23	0.91075	1.48	0.88575	1.73	0.91467	1.98	0.99171
1.24	0.90852	1.49	0.88595	1.74	0.91683	1.99	0.99581
						2.00	1.00000

$$\Gamma(n) = \int_0^\infty t^{n-1} e^{-t} dt$$

$$\Gamma(n) = (n-1)\Gamma(n-1)$$

$$\Gamma(1) = 1$$

$$\Gamma(1/2) = \sqrt{\pi}$$

Source: Suhir, E., *Applied Probability for Engineers and Scientists*, McGraw-Hill, New York, 1997, p. 555. (Reprinted with permission of McGraw-Hill.)

Appendix 8:
Median Ranks Table

Median Ranks

$j\backslash n$	1	2	3	4	5	6	7	8	9	10
1	50.000	29.289	20.630	15.910	12.945	10.910	9.428	8.300	7.412	6.697
2		70.711	50.000	38.573	31.381	26.445	22.849	20.113	17.962	16.226
3			79.370	61.427	50.000	42.141	36.412	32.052	28.624	25.857
4				84.090	68.619	57.859	50.000	44.015	39.308	35.510
5					87.055	73.555	63.588	55.984	50.000	45.169
6						89.090	77.151	67.948	60.691	54.831
7							90.572	79.887	71.376	64.490
8								91.700	82.038	74.142
9									92.587	83.774
10										93.303

Sample Size

Median Ranks

$j\backslash n$	11	12	13	14	15	16	17	18	19	20
1	6.107	5.613	5.192	4.830	4.516	4.240	3.995	3.778	3.582	3.406
2	14.796	13.598	12.579	11.702	10.940	10.270	9.678	9.151	8.677	8.251
3	23.578	21.669	20.045	18.647	17.432	16.365	15.422	14.581	13.827	13.147
4	32.380	29.758	27.528	25.608	23.939	22.474	21.178	20.024	18.988	18.055
5	41.189	37.853	35.016	32.575	30.452	28.589	26.940	25.471	24.154	22.967
6	50.000	45.951	42.508	39.544	36.967	34.705	32.704	30.921	29.322	27.880
7	58.811	54.049	50.000	46.515	43.483	40.823	38.469	36.371	34.491	32.795
8	67.620	62.147	57.492	53.485	50.000	46.941	44.234	41.823	39.660	37.710
9	76.421	70.242	64.984	60.456	56.517	53.059	50.000	47.274	44.830	42.626
10	85.204	78.331	72.472	67.425	63.033	59.177	55.766	52.726	50.000	47.542
11	93.893	86.402	79.955	74.392	69.548	65.295	61.531	58.177	55.170	52.458
12		94.387	87.421	81.353	76.061	71.411	67.296	63.629	60.340	57.374
13			94.808	88.298	82.568	77.525	73.060	69.079	65.509	62.289
14				95.169	89.060	83.635	78.821	74.529	70.678	67.205
15					95.484	89.730	84.578	79.976	75.846	72.119
16						95.760	90.322	85.419	81.011	77.033
17							96.005	90.849	86.173	81.945
18								96.222	91.322	86.853
19									96.418	91.749
20										96.594

Sample Size

Source: Kapur, K.C. and Lamberson, L.R., *Reliability in Engineering Design*, John Wiley, New York, 1977. (Reprinted with permission of John Wiley & Sons, Inc.)

Appendix 9:
Five Percent Ranks Table

Five Percent Ranks

$j\backslash n$	1	2	3	4	5	6	7	8	9	10
					Sample Size					
1	5.000	2.532	1.695	1.274	1.021	0.851	0.730	0.639	0.568	0.512
2		22.361	13.535	9.761	7.644	6.285	5.337	4.639	4.102	3.677
3			36.840	24.860	18.925	15.316	12.876	11.111	9.775	8.726
4				47.237	34.259	27.134	22.532	19.290	16.875	15.003
5					54.928	41.820	34.126	28.924	25.137	22.244
6						60.696	47.930	40.031	34.494	30.354
7							65.184	52.932	45.036	39.338
8								68.766	57.086	49.310
9									71.687	60.584
10										74.113

Five Percent Ranks

$j\backslash n$	11	12	13	14	15	16	17	18	19	20
					Sample Size					
1	0.465	0.426	0.394	0.366	0.341	0.320	0.301	0.285	0.270	0.256
2	3.332	3.046	2.805	2.600	2.423	2.268	2.132	2.011	1.903	1.806
3	7.882	7.187	6.605	6.110	5.685	5.315	4.990	4.702	4.446	4.217
4	13.507	12.285	11.267	10.405	9.666	9.025	8.464	7.969	7.529	7.135
5	19.958	18.102	16.566	15.272	14.166	13.211	12.377	11.643	10.991	10.408
6	27.125	24.530	22.395	20.607	19.086	17.777	16.636	15.634	14.747	13.955
7	34.981	31.524	28.705	26.358	24.373	22.669	21.191	19.895	18.750	17.731
8	43.563	39.086	35.480	32.503	29.999	27.860	26.011	24.396	22.972	21.707
9	52.991	47.267	42.738	39.041	35.956	33.337	31.083	29.120	27.395	25.865
10	63.564	56.189	50.535	45.999	42.256	39.101	36.401	34.060	32.009	30.195
11	76.160	66.132	58.990	53.434	48.925	45.165	41.970	39.215	36.811	34.693
12		77.908	68.366	61.461	56.022	51.560	47.808	44.595	41.806	39.358
13			79.418	70.327	63.656	58.343	53.945	50.217	47.003	44.197
14				80.736	72.060	65.617	60.436	56.112	52.420	49.218
15					81.896	73.604	67.381	62.332	58.088	54.442
16						82.925	74.988	68.974	64.057	59.897
17							83.843	76.234	70.420	65.634
18								84.668	77.363	71.738
19									85.413	78.389
20										86.089

Source: Kapur, K.C. and Lamberson, L.R., *Reliability in Engineering Design*, John Wiley, New York, 1977. (Reprinted with permission of John Wiley & Sons, Inc.)

Appendix 10: Ninety-Five Percent Ranks Table

Ninety-Five Percent Ranks

					Sample Size					
$j\backslash n$	1	2	3	4	5	6	7	8	9	10
1	95.000	77.639	63.160	52.713	45.072	39.304	34.816	31.234	28.313	25.887
2		97.468	86.465	75.139	65.741	58.180	52.070	47.068	42.914	39.416
3			98.305	90.239	81.075	72.866	65.874	59.969	54.964	50.690
4				98.726	92.356	84.684	77.468	71.076	65.506	60.662
5					98.979	93.715	87.124	80.710	74.863	69.646
6						99.149	94.662	88.889	83.125	77.756
7							99.270	95.361	90.225	84.997
8								99.361	95.898	91.274
9									99.432	96.323
10										99.488

Ninety-Five Percent Ranks

					Sample Size					
$j\backslash n$	11	12	13	14	15	16	17	18	19	20
1	23.840	22.092	20.582	19.264	18.104	17.075	16.157	15.332	14.587	13.911
2	36.436	33.868	31.634	29.673	27.940	26.396	25.012	23.766	22.637	21.611
3	47.009	43.811	41.010	38.539	36.344	34.383	32.619	31.026	29.580	28.262
4	56.437	52.733	49.465	46.566	43.978	41.657	39.564	37.668	35.943	34.366
5	65.019	60.914	57.262	54.000	51.075	48.440	46.055	43.888	41.912	40.103
6	72.875	68.476	64.520	60.928	57.744	54.835	52.192	49.783	47.580	45.558
7	80.042	75.470	71.295	67.497	64.043	60.899	58.029	55.404	52.997	50.782
8	86.492	81.898	77.604	73.641	70.001	66.663	63.599	60.784	58.194	55.803
9	92.118	87.715	83.434	79.393	75.627	72.140	68.917	65.940	63.188	60.641
10	96.668	92.813	88.733	84.728	80.913	77.331	73.989	70.880	67.991	65.307
11	99.535	96.954	93.395	89.595	85.834	82.223	78.809	75.604	72.605	69.805
12		99.573	97.195	93.890	90.334	86.789	83.364	80.105	77.028	74.135
13			99.606	97.400	94.315	90.975	87.623	84.366	81.250	78.293
14				99.634	97.577	94.685	91.535	88.357	85.253	82.269
15					99.659	97.732	95.010	92.030	89.009	86.045
16						99.680	97.868	95.297	92.471	89.592
17							99.699	97.989	95.553	92.865
18								99.715	98.097	95.783
19									99.730	98.193
20										99.744

Source: Kapur, K.C. and Lamberson, L.R., *Reliability in Engineering Design*, John Wiley, New York, 1977. (Reprinted with permission of John Wiley & Sons, Inc.)

Appendix 11:
Critical Values for the
Kolmogorov–Smirnov
Statistic (d_α)

Critical Values for the Kolmogorov–Smirnov Statistic (d_α)

Sample Size (n)	Level of Significance (α)				
	0.20	0.10	0.05	0.02	0.01
1	0.900	0.950	0.975	0.990	0.995
2	0.684	0.776	0.842	0.900	0.929
3	0.565	0.636	0.708	0.785	0.829
4	0.493	0.565	0.624	0.689	0.734
5	0.447	0.509	0.563	0.627	0.669
6	0.410	0.468	0.519	0.577	0.617
7	0.381	0.436	0.483	0.538	0.576
8	0.358	0.410	0.454	0.507	0.542
9	0.339	0.387	0.430	0.480	0.513
10	0.323	0.369	0.409	0.457	0.489
11	0.308	0.352	0.391	0.437	0.468
12	0.296	0.338	0.375	0.419	0.449
13	0.285	0.325	0.361	0.404	0.432
14	0.275	0.314	0.349	0.390	0.418
15	0.266	0.304	0.338	0.377	0.404
16	0.258	0.295	0.327	0.366	0.392
17	0.250	0.286	0.318	0.355	0.381
18	0.244	0.279	0.309	0.346	0.371
19	0.237	0.271	0.301	0.337	0.361
20	0.232	0.265	0.294	0.329	0.352
21	0.226	0.259	0.287	0.321	0.344
22	0.221	0.253	0.281	0.314	0.337
23	0.216	0.247	0.275	0.307	0.330
24	0.212	0.242	0.269	0.301	0.323
25	0.208	0.238	0.264	0.295	0.317
26	0.204	0.233	0.259	0.290	0.311
27	0.200	0.229	0.254	0.284	0.305
28	0.197	0.225	0.250	0.279	0.300
29	0.193	0.221	0.246	0.275	0.295
30	0.190	0.218	0.242	0.270	0.290
31	0.187	0.214	0.238	0.266	0.285
32	0.184	0.211	0.234	0.262	0.281
33	0.182	0.208	0.231	0.258	0.277
34	0.179	0.205	0.227	0.254	0.273
35	0.177	0.202	0.224	0.251	0.269
36	0.174	0.199	0.221	0.247	0.265
37	0.172	0.196	0.218	0.244	0.262
38	0.170	0.194	0.215	0.241	0.258
39	0.168	0.191	0.213	0.238	0.255
40	0.165	0.189	0.210	0.235	0.252
Over 40	$1.07/\sqrt{n}$	$1.22/\sqrt{n}$	$1.36/\sqrt{n}$	$1.52/\sqrt{n}$	$1.63/\sqrt{n}$

Source: Sheskin, D.J., *Handbook of Parametric and Nonparametric Statistical Procedures*, 3rd ed., Chapman & Hall/CRC, London, 2004. (Reprinted with permission of Chapman & Hall/CRC.)

Appendix 12: Answers to Problems

CHAPTER 2: COMPONENT REPLACEMENT DECISIONS

To be solved using the mathematical models.

1. $C(1) = (0 + 1200)/1 = \$1200/\text{month}$
   $C(2) = (300 + 1200)/2 = \$750/\text{month}$
   $C(3) = \$700/\text{month (minimum)}$
   $C(4) = \$800/\text{month}$
2. $C(10K) = \$14.30/1000 \text{ km}$
   $C(20K) = \$10.00/1000 \text{ km}$
   $C(30K) = \$9.33/1000 \text{ km (minimum)}$
   $C(40K) = \$10.00/1000 \text{ km}$
3. $C(t_p = 5K) = \$0.093/\text{km}$
   $C(10K) = \$0.067/\text{km}$
   $C(15K) = \$0.063/\text{km (minimum)}$
   $C(20K) = \$0.078/\text{km}$
4. $D(2) = 0.94 \text{ day/month}$
   $D(4) = 0.77 \text{ day/month (minimum)}$
   $D(6) = 0.78 \text{ day/month}$
   $D(8) = 0.88 \text{ day/month}$
5. $M(t_p = 5K) = 2.5K$, $M(t_p = 10K) = 5.0K$, etc.
   $D(t_p = 5K) = 1.03 \text{ days/1000 km}$
   $D(t_p = 10K) = 0.80 \text{ day/1000 km}$
   $D(t_p = 15K) = 0.80 \text{ day/1000 km}$
   $D(t_p = 20K) = 0.90 \text{ day/1000 km}$

To be solved using Glasser's graphs.

6. $t_p = 31,148$, $\rho = 92\%$
7. a. $t_p = 60.7 \text{ hours}$, $\rho = 36\%$
   b. $t_p = 58.1 \text{ hours}$, $\rho = 38\%$
8. $t_p = 17,900 \text{ km}$, savings $= 40\%$
9. a. $t_p = 117,000 \text{ km}$
   b. savings $= 86\%$
   c. R-o-o-F cost $= \$2000/150,000 \text{ km} = \$0.0133/\text{km}$
   Optimal policy cost $= 0.14 \, (0.0133) = \$0.0018/\text{km}$

10. $z = -2.3$, $t_p = 177$ hours, $\rho = 83\%$
11. $z = -2.2$, $t_p = 117$ hours, $\rho = 45\%$

To be solved using the OREST software.

12. a. $\beta = 2.42$, $\eta = 19.00$ weeks, mean life = 16.84 weeks
    b. Preventive replacement at 6.66 weeks

Age Weeks	Cost $/Week	% Failure Replacement
3	38.92	1
4	30.31	2
5	27.69	4
6	26.04	6
7	25.86	8
8	26.44	12

13. a. Ages (km) at failure are 51,220, 16,840, 45,380, and 58,130. Suspensions occur at 47,620 and 29,210. These are obtained by subtracting odometer readings from the next higher reading.
    b. $\beta = 2.16$, $\eta = 54,745$, mean life = 48,483 km
    c. Wear-out
    d. Maybe curved, getting steeper, more serious wear-out possible
    e. 18,612 km, 0.01 $/km, 0.01 $/km
    f. For 20,000 km, $0.01 $/km
    g. Utilization = $30 \times 50,000 = 1,500,000$ km/year, spares = 84
    h. For preventive replacement at 20,000 km, spares = 78
14. a. $\beta = 0.36$, $\eta = 31.21$ hours, mean life = 136.53 hours
    b. No. Replace only on failure. More cautiously, extend preventive replacement age to, say, 30 hours, and check if there is any wear-out.
    c. 111, 81
15. a. $\beta = 2.26$, $\eta = 234,067$ km, mean life = 207,330 km
    b. 86,598 km
    c. 240,000 km
    d. i. 80,439 km
       ii. 13 replacements, 8.5% failure replacements
    e. i. $2500
       ii. $500
16. Wear-out, $\beta = 1.64$
    a. Design defect
    b. No

CHAPTER 3: INSPECTION DECISIONS

1. a. $D(n) = \lambda(n) T_f + nT_i$
    b. n = ¼, availability = 93.8%, with k = 1/32
2. a. $D(n) = (\lambda_1(n) + \lambda_2(n))T_f + nT_i$
    b. $n = 1$, A = 0.9
    c. $K_1 = 0.4$
3. a. FFI < 3.46 hours
    b. Not available
    c. A > $1 - 2 \times 10^{-7}$
4. a. $h = 3.43 \times 10^{-5}$, n = 1
       $h = 3.35 \times 10^{-5}$, n = 2
       $h = 1.87 \times 10^{-5}$, n = 3
    b. Not available
    c. Not available
5. a. CM data and Event data
    b. Not available
    c. $h = 0.00202$, n = 1
       $h = 0.000009$, n = 2
       $h = 0.00063$, n = 3
    d. Not available

CHAPTER 4: CAPITAL EQUIPMENT REPLACEMENT DECISIONS

To be solved using the mathematical models.

1. $C(1) = \$4500$ $C(5) = \$3973$ (minimum)
    $C(2) = \$4225$ $C(6) = \$3982$
    $C(3) = \$4065$ $C(7) = \$4045$
    $C(4) = \$3989$ $C(8) = \$4146$
2. $n = 6$ optimal
3. $r = 0.91$
    EAC(1) = \$16,698
    EAC(2) = \$10,558
    EAC(3) = \$8459
    EAC(4) = \$7177
    EAC(5) = \$6824 (minimum)
    EAC(6) = \$7244
    EAC(7) = \$8072
4. $C(1) = \$110,561$
    $C(2) = \$75,639$
    $C(3) = \$68,442$
    $C(4) = \$67,848$ (optimal)

5. Using model:

$$\text{EAC}(n) = \left(A + \sum_{i=1}^{n} C_i r^i - R_n r^n \right) \times \text{CRF}(n)$$

$C(1) = \$90{,}981$ (optimal)
$C(2) = \$100{,}344$
$C(3) = \$117{,}274$
$C(4) = \$129{,}872$
$C(5) = \$139{,}866$

To be solved using the educational versions of the AGE/CON or PERDEC software.

6. Year 3, EAC = $79,973
7. Year 3, EAC = $17,461
8. Year 2, EAC = $21,500
9. Year 2, EAC = $26,052
10. Year 3, EAC = $14,926
11. When utilization = 10,000 miles, economic life is 4 years with EAC = $35,189.
    When utilization = 8000 miles, economic life is 4 years with EAC = $30,068.
12. Year 2, EAC = $30,398
13. Year 3, EAC = $33,593
14. Using linear trend $Y = 1.40708 + 6.63497e^{-5}x$, the economic life is 3 years, EAC = $29,925.
    When using a polynomial of order 3, the economic life is 2 years, EAC = $28,312.
15. Year 4, EAC = $14,235
16. With 16% interest rate, Year 3, EAC = $18,134
    With 19% interest rate, Year 3, EAC = $18,524

CHAPTER 5: MAINTENANCE RESOURCE REQUIREMENTS

1. $n = 6$ optimal $C(n) = nC_l + W_s \lambda C_d$

2. a. $C(n) = n \times C_w + \int_0^\infty T(r) \times f(r)\,dr,$

   where

   $$T(r) = \begin{cases} C_r \times r & , \quad r \le nm \\ C_r \times nm + \min\left[(r-nm)C_1, (r-nm)C_2(r-nm) \right], & r > nm \end{cases}$$

   b. $n = 6$ optimal

APPENDIX 1: STATISTICS PRIMER

1. $R(104) = 0.3446$, $h(105) = 0.1141$ per hour
2. $R(5) = 0.9878$, $R(25) = 0.9405$
3. $R(100) = 0.99$, $h(100) = 0.0002$ per hour
4. $P(T > 1220|T > 1200) = 0.975$
5. $R(4100) = 0.8159$, $h(4400) = 0.000459$ failures per hour
6. and 7. These problems involve finding the form of functions and sketching them. No answer is provided.

APPENDIX 2: WEIBULL ANALYSIS

1. $\beta = 1.5$, $\eta = 25,000$ miles, $\mu = 21,000$ miles
2. $\beta = 2.0$, $\eta = 68,000$ miles, $\mu = 60,000$ miles
3. $\beta = 1.5$, $\eta = 94,000$ miles, $\mu = 85,000$ miles
4. $\beta = 1.3$, $\mu = 95,000$ miles
5. $r(t)$ decreases to about 1000 hours, then it remains constant. A high rate of manufacturer's built-in failures is possible. This indicates a high rate of manufacturing defects.
6. $\beta = 1.82$, $\eta = 9000$ cycles, $R(1000|3000) = 90.8\%$
7. $\beta = 2$, $\eta = 22.6$ months, acceptable according to K-S test
8. a. $\beta = 1.54$, shape factor; $\eta = 4669$, characteristic life
   b. $\mu = 4202$ hours, $\sigma = 2782$
   c. Yes, $d < d_\alpha$
   d. 40%
   e. 1080 hours
   f. $\gamma = 1097$, failure-free period
   g. $\mu = 4246$ hours, $\sigma = 3189$ hours
9. a. 9; 136 is a suspension
   b. $d = 0.208$, $d_\alpha = 0.388$, the hypothesis is not rejected
   c. $\beta = 10.6$, $\eta = 355$, $\gamma = -181$, $\mu = 157$, $\sigma = 38.6$
10. a. $\gamma = 169$, the failure free period. For $t = 5000$ hours, $t - \gamma = 4831$ hours; $F(t - \gamma) = 53.2\%$
    b. For $F(t - \gamma) = 20\%$, point estimate of $t = 2789$ hours, 90% confidence interval of t is (1469, 4399) hours
11. a. Test statistic of Laplace trend test, $u = 1.2 < 1.96$, critical value of u at $\alpha = 5\%$. Thus, no trend in the time between failures is detected. Probability plot can be used to model the time-between-failures distribution.
    b. $\beta = 2.51$, $\eta = 8679$, $\gamma = 0$, $\mu = 7701$, $\sigma = 3383$
    From the Weibull plot, $R(5000) = 72\%$. Thus, the reliability target at 5000 copies cannot be met.
12. $R(100) = 42\%$
    The 2 failure modes are assumed to be independent of each other.

Index